JN246545

和菓子

WAGASHI

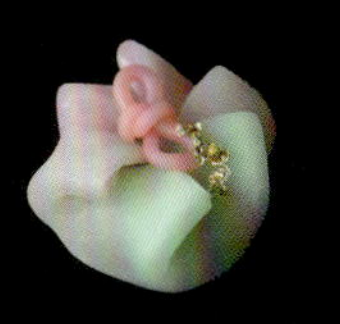

WAGASHI

Author : Hajime Nakamura

Designer: Yukie Kamauchi, Yukihiko Ishikawa(GRiD)
Rayout: Mutsumi Nakanishi
Translator : Ruth S. McCreery (The Word Works)
Editor : Masanori Omae (Lampoon House)

Publisher : Masaru Onodera
Published by KAWADE SHOBO SHINSHA Ltd.Publishers
2-32-2 Sendagaya, Shibuya-ku, Tokyo 151-0051, Japan
+81-3-3404-1201(Sales department)
+81-3-3404-8611(Editorial department)
http://www.kawade.co.jp/

First edition : January 30, 2013
Second edition: July 30, 2014
This edition: January 30, 2018
Printed & Bound by Dai Nippon Printing Co.,Ltd.

和菓子

WAGASHI

Hajime Nakamura

中村 肇

目　次 / Contents

はじめに

和菓子……それも生菓子に出会った時に、それまで意識してこなかった、
生まれ育った京都の色や形が見え、そして息づかいが聞こえて来ました。
生菓子を買い求め、その生菓子から想像される風景を捜し求めて写真に収め、
ブログで紹介し始めて気がつくと7年の歳月が経っていました。

和菓子にはその色彩、造形的な工夫やそれぞれに付けられた「菓子の名前」など、
どこを取っても隙間なく美への思いが凝縮されています。

あたたかいお茶を飲みながら、和菓子の楽しみが本書を通じて伝われば幸いです。

2013年1月吉日

中村　肇

A wealth of *wagashi*, Japanese confectionary, inspired by the everyday events and customs, the seasonal flowers and festivals of Kyoto: Only someone born and raised in Kyoto could write this book. I hope you will enjoy, through these beautiful photographs and the stories that accompany them, the utter delight that I take in my favorite sweets as an intimate part of daily life in Kyoto.

Hajime Nakamura

和 菓 子

和菓子の最も特徴的なところは、その季節感や風物詩にあるといえます。古代からさまざまな工夫を凝らし、日本の風土や季節に合わせて独自のスタイルを創りだしてきました。一般的には日本の伝統的な製法で作られた菓子のことを和菓子と呼びます。明治(時代)以降にヨーロッパから入ってきた菓子は洋菓子、それ以前に大陸からもたらされた唐菓子や南蛮渡来の菓子は和菓子とされています。

和菓子はお茶席において、鑑賞する楽しみと、味わう楽しみを兼ね備えたものとして、工夫され発達してきました。お薄茶席では干菓子、濃茶席では生菓子が出され、共にいただきます。砂糖が材料に使われだすのは近世以降で、それ以前は独特の風味と甘さの和三盆が、和菓子の発達に大きな役割を担ってきました。

洋菓子のように生のくだものを使うことは少なく、砂糖、水飴、米、小麦、小豆などを使用し、油はほとんど使いません。含まれる水分量によって干菓子(乾菓子)、生菓子または半生菓子と分類されます。

菓子職人はそれぞれの素材の特徴を生かして季節感をいかに表現するかを競い、夏の涼感を表現するのに葛を使って水の透明感を出したり、祭の衣裳を精緻な技術で描きだしたりしました。

Wagashi: Japanese Confectionary

The term *wagashi*, for confectionary made using methods traditional to Japan, was coined after the introduction of European culinary traditions in the mid nineteenth century. It has become a term that also includes the pastries brought to Japan by its envoys to China (*karagashi*) and those later introduced by the Spanish and Portuguese (*nanbangashi*).

The sugar-based *wagashi* served in the tea ceremony, with both thin and thick tea (*usucha* and *koicha*), are as beautiful as they are delicious. Moist sweets are usually served with thick tea and dry sweets with thin tea.

Because they are served with tea, these *wagashi* are often quite sweet and are made with almost no oil. The short list of ingredients includes sugar, *mizuame* (a sweetener made by converting rice or sweet potato starch to syrup), rice, wheat flour, and *azuki* beans. Poached or dried fruit is also used, but almost no fresh fruit.

The use of sugar as a *wagashi* ingredient dates from the early modern period. In particular, in the Edo period, when white sugar was hard to come by, refined Japanese sugar (*wasanbon*), with its distinctive flavor and nicely balanced sweetness, contributed to dramatic advances in *wagashi*. Given that the sweetest treats, before sugar became available, were persimmons, it is possible to imagine the delicacy of the sweetness of *wagashi*. These confections were required to be as delightful to look at as they were delicious. The ingredients were also chosen to be seasonally appropriate. In the summer, for example, *wagashi* might be given a translucent coating, to communicate a refreshing sense of coolness. In confectionary raised to the level of an art form, these edible delights expressed the wonders of nature.

Wagashi can be broadly grouped into three types, according to moisture content. Those with a moisture content of 20 percent or less are referred to as dry sweets (*higashi*); those with 40 percent or more are moist sweets (*namagashi*), and those in between are semi-moist (*han namagashi*).

京菓子 / Kyoto Confectionary

京都では、宮中や公家、寺社、茶家などに納めるための「上菓子」として、特別な祝いや祭のための洗練された意匠が施され、独自の発達を遂げました。それと同時に「おまん（饅頭）」「だんご」など日常に食する菓子を年中行事に合わせてつくるお店は「おまんやさん」「おもちやさん」と呼ばれ、生活の中に浸透していきました。今でも、日常食としておはぎやおもちが、うどんや寿司などと一緒に売られているお店が結構あります。

Kyoto confectionary includes the deluxe art confectionary (*jogashi*) created on special order for members of the nobility, temples and shrines, tea masters, and for special celebrations, together with everyday types of *o-kashi*, including dumplings (*o-man, dango*) and sweets made of glutinous rice. The creators of deluxe *jogashi* confectionary are designated *kashi tsukasa*, confectionary artists. Ordinary confectionary makers are simply referred to by the type of *wagashi* they produce.

Deluxe confectionary became increasingly sophisticated as it evolved for use as gifts and offerings and for the tea ceremony. Ordinary confectionary, as treats to be consumed during the seasonal events held throughout the year, also became increasingly diverse.

上菓子

こなし

白こしあん（手亡豆（てぼうまめ）などの隠元豆（いんげん）、あるいは白小豆（しろあずき）のあん）と薄力粉を混ぜて蒸したものに砂糖水を加え、練り上げたもの。色をつけてさまざまな形に加工する。梅の蕾をかたどった「未開紅」、紅葉に仕立てた「竜田川」をはじめ、葛菓子のあんなどさまざまな表現に使われます。

きんとん

蒸した山芋を裏漉して砂糖と炊いたもの（薯蕷（じょうよ）煉切り）や、白あんを寒天で固めたもの（きんとんあん、天餡）、白あんを求肥でつないだもの（煉切り）を、いろいろな色にそめ、裏漉し器でそぼろ状にし、あんなどの芯に植えつけて季節を表現します。

求 肥（ぎゅうひ）

もち米の粉を水で練って湯がき、火の上で砂糖を加えて練ったもの。夏の菓子「鮎」、「調布」などに使われます。

く ず

本葛粉に水を加えたものを漉して、砂糖を加え加熱し、アルファ化させたもの。葛切り、葛饅頭などは、その透明感が涼しさを呼ぶ。また六方を焼いただけの「葛焼」は熟練を要する菓子。

薯 蕷（じょうよ）

山芋のこと。「織部まんじゅう」など上用饅頭は、山芋をすり下ろして砂糖と上用粉（細目の米粉）を加えたものであんを包み、蒸して作ります。また、すり下ろした山芋に、砂糖、水、軽羹粉（粗目の米粉）を加え、蒸し上げたものがカルカン（軽羹）。蒸した山芋を裏漉しして砂糖と炊いたものが、薯蕷煉切り。

この他にも中間素材は数多くあります。

それぞれの素材の味を引き出す技術が必要で、さらに季節感や供される場所の状況やコンセプトに応じた意匠が、菓子店や菓子職人の感性で表現されます。

Deluxe *Jogashi* Confectionary

These beautiful sweets mainly use the ingredients and methods described below in brilliant examples of the art of confectionary.

Konashi This highly plastic material is made of a steamed strained bean paste and cake flour mixture that is kneaded with sugar water. It can be colored and sculpted into a variety of shapes, from plum flower buds to autumn leaves.

Kinton Kinton, which may be of several types (made of strained steamed yams boiled with sugar, white bean paste thickened up with agar-agar, or white bean paste bound with *gyuhi*, below), is dyed various colors, sieved to create slender threads or beads, and arranged atop bean paste or other fillings.

Gyuhi Glutinous rice flour is kneaded with water, parboiled, sugar added while on the heat, and kneaded again. This soft mixture is used in summer sweets such as *ayu* and *chofu*.

Kuzu Starch made from the root of the kudzu vine is dissolved in water, strained, sugar added, and heated until it jells. Its translucency gives confections a refreshingly cool appearance.

Joyo *Joyo*,another ingredient based on grated yams, is used to make the outer layer of superb *joyo* dumplings. Sugar and fine rice flour are added to grated yams. The resulting paste is wrapped around bean paste and the dumplings steamed. Sugar, water, and coarse rice flower may also be added to the grated yams and steamed to make the dumplings known as *karukan*. Yams may also be sieved and heated with sugar to make the delicate sweets known as *joyo nerikiri*. Whatever the technique, the point is to make effective use of the whiteness intrinsic to yams and of their distinctive flavor.

Semi-processed materials, prepared by boiling, steaming, mixing, kneading, or otherwise processing ingredients, are also used in creating *jogashi*. The ingredients for these beautiful confections must be carefully selected and handled with skill to draw out their flavor, with each of the steps in their preparation carried out properly. The art of *jogashi* also requires expressing a sense of the season or a concept of the place in which the confection will be enjoyed. Creating deluxe *jogashi* thus requires a delicate artistic sensibility as well as highly trained skills. With the many confectionary shops and craftsmen applying the techniques of their specific traditions and the skills they have honed themselves, while expressing their own sensibilities, the result is a wealth of subtly different, distinctive *wagashi*.

和菓子の歴史

食物が充分でなかった古代の人々は、ときには野生の「古能美」(木の実)や「久多毛能」(果物)を採集して空腹を満たしていました。いまでも果物のことを「水菓子」と呼ぶように、菓子は本来、木の実や果物を表わす言葉でした。しかし材料の面で見てみると、今日和菓子と呼ぶものの原形は、餅や団子からきていると考えていいでしょう。

縄文時代晩期には大陸から水稲耕作が伝わり、農耕を中心とした生活に変わりましたが、まだ食物が充分とはいえませんでした。そこで、木の実を石臼や石槌で砕いて粉にし、水に晒して灰汁(あく)抜きをしたものをだんご状に丸め、熱を加えるなどして保存食としていました。これらは主食というより間食で、ときには植物の蜜や果物の汁などで甘味をつけて味わっていたとも考えられ、これが餅や団子の始まりといわれています。そしてその一方で、外国から新しい外来の菓子が伝わることにより、日本の菓子の歴史に変化が生じることになります。

History of *Wagashi*

In prehistoric times, people supplemented their diets with wild fruit or berries. The connection between snacks and fruit is probably why the term *kashi* ("fruit of the tree") has come to refer to sweet snacks, as in *o-kashi* or *wagashi*. To the ancients, the sweetness of fruit must have been a delight, and such snacks were distinguished from their main diet.

While agriculture had been introduced, the food supply was still often insufficient, and people supplemented their diets with acorns. Acorns are, however, too bitter to eat as is. Thus, people would crush acorns, soak the meal in water to leach out the bitterness, then form the resulting paste into balls and cook them. That was the origin of *dango*, small balls formed of various edible ingredients, that are now a classic type of *wagashi*.

In time, *mochi* (a pliable dough made by soaking, steaming, and pounding glutinous rice), now thought to be Japan's oldest processed food, was born. The *Wamyo ruiju sho*, a Chinese-Japanese dictionary dating from 934, has entries concerning glutinous rice and dishes made from it. Because they were made of rice, the most precious of ingredients, such foods were regarded as sacred, as such ancient documents as the *Bungo fudoki*, an eighth-century gazetteer, make clear.

唐菓子

外来の菓子には、遣唐使がもたらした「唐菓子」、鎌倉から室町時代にかけて中国（宋・元）に留学した禅僧とともに渡来した羊羹（ようかん）や饅頭などの「点心」の菓子、安土桃山時代にポルトガルやスペインの宣教師が伝えた「南蛮菓子」があります。

奈良時代に中国から渡来した唐菓子には、梅子（ばいし）・桃子（とうし）・桂子（けいし）・団喜（だんき）・椿餅（つばいもち）・糫餅（まがりもち）・餛飩（こんとん）、煎餅（いりもち）などの唐菓子と果餅がありました。果餅は米、麦、大豆、小豆などをこねたり、油で揚げたりしたもので、特徴のある形をしており祭祀用に尊ばれました。

Karagashi

When Japan began sending envoys to China (19 times, between 630 and 894), one of the things they brought back was "Chinese confectionary," or *karagashi*. These exotic sweets were made of rice, wheat, soybeans, *azuki* beans, and other ingredients, kneaded together and deep-fried in special shapes. They were treasured for festival use. The importation of *karagashi* is thought to have had a major impact on the evolution of *wagashi*.

茶席のお菓子

茶席には、「点心」という食事以外の軽食がありました。そのなかに「羹」という汁があり、猪羹、白魚羹、芋羹、鶏鮮羹など種類も多く、「羊羹」と呼ばれるものもありました。羊羹は本来、羊の肉の入った汁でしたが、獣肉食は公然ではなかった日本では、羊の肉に似せて小麦や小豆の粉を葛粉に混ぜて蒸し固めたものを入れました。そしてその羊の肉に似せたものが汁物から離れて誕生したのが「羊羹」の始まりで、当時は「蒸羊羹」と呼んでいました。

Wagashi and the Tea Ceremony

Another major influence on *wagashi* was the fashion for drinking tea and the development of the tea ceremony. Tea is said to have been brought to Japan from China by the Zen priest Eisai in about 1191; enjoyment of it spread gradually. At gatherings to enjoy tea in the Muromachi period (1392 – 1573), it was customary to serve a hot soup, or *atsumono*, of which there were forty-eight types. One of those was *yokan*, a hot soup based on mutton, the eating meat is non habit in Japan; as a substitute, ingenious chefs developed imitations of mutton made of a mixture of wheat and *azuki* bean flour. Over time, the dish evolved away from its origins as soup into the translucent confectionary we know today as *yokan*.

The discovery of agar-agar and the conversion of that *azuki* bean mixture to the jelly-like sweet we known as *neri yokan* occurred in about 1800. Other confections served in the tea ceremony also contributed to the evolution of *wagashi*.

南蛮菓子

南蛮菓子は現在でも食べられている和菓子の原型で、ボーロ、カステイラ（カステラ）、金平糖（こんぺいとう）、ビスカウト（ビスケット）、パン、有平糖（ありへいとう）、鶏卵素麺などがあります。

南蛮菓子が到来した時期が、千利休によりわび茶が大成された茶の湯の興隆期にあったこともあって、和菓子はこれらの外来菓子に大きな影響を受けながら独自の発展を遂げるのです。

初期の茶会の菓子は現代のものとは趣が違い、栗や榧（かや）などの木の実、柿などの果物、昆布、餅や饅頭などでした。当時は砂糖はまだ高価な輸入品で、菓子には入れず添えられる場合が多かったようです。また饅頭といっても現在のような甘い小豆入りのものではなかったと考えられます。

Nanbangashi

Spanish and Portuguese visitors reaching Japan in the sixteenth century brought with them *nanbangashi*, the ancestral forms of types of *wagashi* enjoyed today, including the Portuguese *bolo*, *castella*, *fios de ovos* (*keiran somen*, in Japanese), and *confeito* (*conpeito*), as well as bread.

江戸時代から現代へ

江戸時代に入り、徳川幕府のもと、社会が安定し経済が発展すると砂糖の輸入量も増えて、菓子作りを専門とする店もでき始め、その技術も飛躍的に向上しました。この時代、全国の城下町や門前町で独自の和菓子が生まれています。

花鳥風月にちなんだ美しい（銘や意匠の）菓子が京都で生まれ、京菓子は高級菓子としてしだいに評判になりました。この頃、京都の和菓子と東京の上菓子が競い合うようにして、菓銘や意匠に工夫を凝らした和菓子が次々に誕生しました。現在食べられている和菓子の多くは、江戸時代に誕生したものなのです。

From Edo Period to Now

The Edo period marked the end of a long period of civil wars. With peace and prosperity, energy could be put into creating confectionary, which evolved dramatically. Castle towns and communities around major temples came up with their own distinctive *wagashi*, and competition emerged between Kyoto and Edo confectionary, with great ingenuity applied to inventing new names and designs for their creations. Most of the *wagashi* enjoyed today originated in the Edo period.

The rapid introduction of European culture in the Meiji period (1868 – 1912) influenced the evolution of *wagashi*. Western cooking equipment, including the oven, made a particular impact, with the birth of new types of baked confectionaries, including the popular chestnut *kuri manju* and castella *manju*.

Wagashi's development has been stimulated over the centuries through interactions with alien cultures. While enjoying imported confectionary, Japanese have managed not to imitate but to assimilate them, creating brilliant new forms of sweets distinctive to Japan. Japanese creativity came into full play in the evolution of *wagashi*.

南　座 *Minamiza*

松飾り *Matsukazari*

新春
New Year's

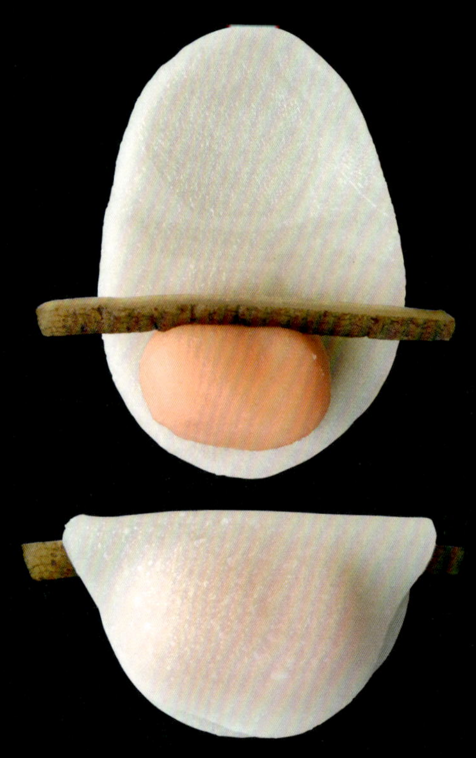

長久堂「花びら餅」　餅皮、みそあん、ごぼう入り
Chokyudo "Hanabiramochi" *Mochikawa, Misoan, Burdock

お正月といえば「花びら餅」、もともとは、平安時代の宮中の新年行事「歯固めの儀式」を簡略化したもの。歯固めの儀式は長寿を願い、餅の上に赤い菱餅を敷き、その上に猪肉や大根、鮎の塩漬け、瓜などをのせて食べていたそうです。それを宮中に菓子を納めていた川端道喜が作っていたそうで、鮎はごぼうで、雑煮は餅と味噌餡で模しています。川端道喜は、文亀三年（1503）創業というので、その頃から花びら餅があったということになります。

New Year's Means *Hanabiramochi*

The Heian court's annual events included a tooth-hardening ritual, in effect a ceremony to ensure long life (for which good teeth are needed). The ritual included eating a diamond-shaped layer of red *mochi*, plus wild boar, giant radish, salted sweetfish, and gourd, served on a layer of *mochi*. The New Year's sweet we know as *hanabiramochi* is thought to be a simplified encapsulation of that ritual, with burdock representing the sweetfish and a bean paste jam the other ingredients, enfolded in a circle of *mochi*. Said to have been invented by Kawabata Doki, who supplied confectionary to the imperial court, *hanabiramochi* has existed since he opened his confectionary business in 1503.

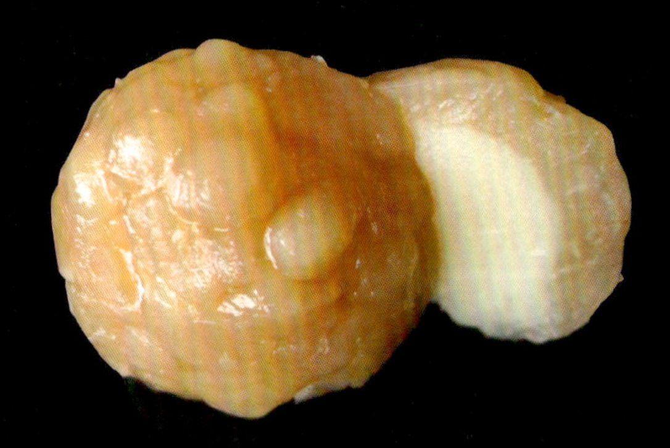

塩芳軒「初日の出」 白小豆、白あん
Shioyoshiken "Hatsuhinode" *Shiroazuki, Shiroan

冬日、梅の蕾も固いけど準備を始めていました。
On a winter day, the plum flower buds were still tight, but preparations were underway.

北野天満宮　Kitano Tenman-gu Shrine

「十牛図」とは、逃げ出した牛を探し求める牧人を喩えとして、牛、すなわち真実の自己を究明する禅の修行によって高まりゆく心境を十段階で示したものです。

The *Jugyu no Zu* (Ten Ox-Herding Pictures) illustrate, through the metaphor of a herdsman searching for a lost ox, the ten stages towards enlightenment in Zen practice.

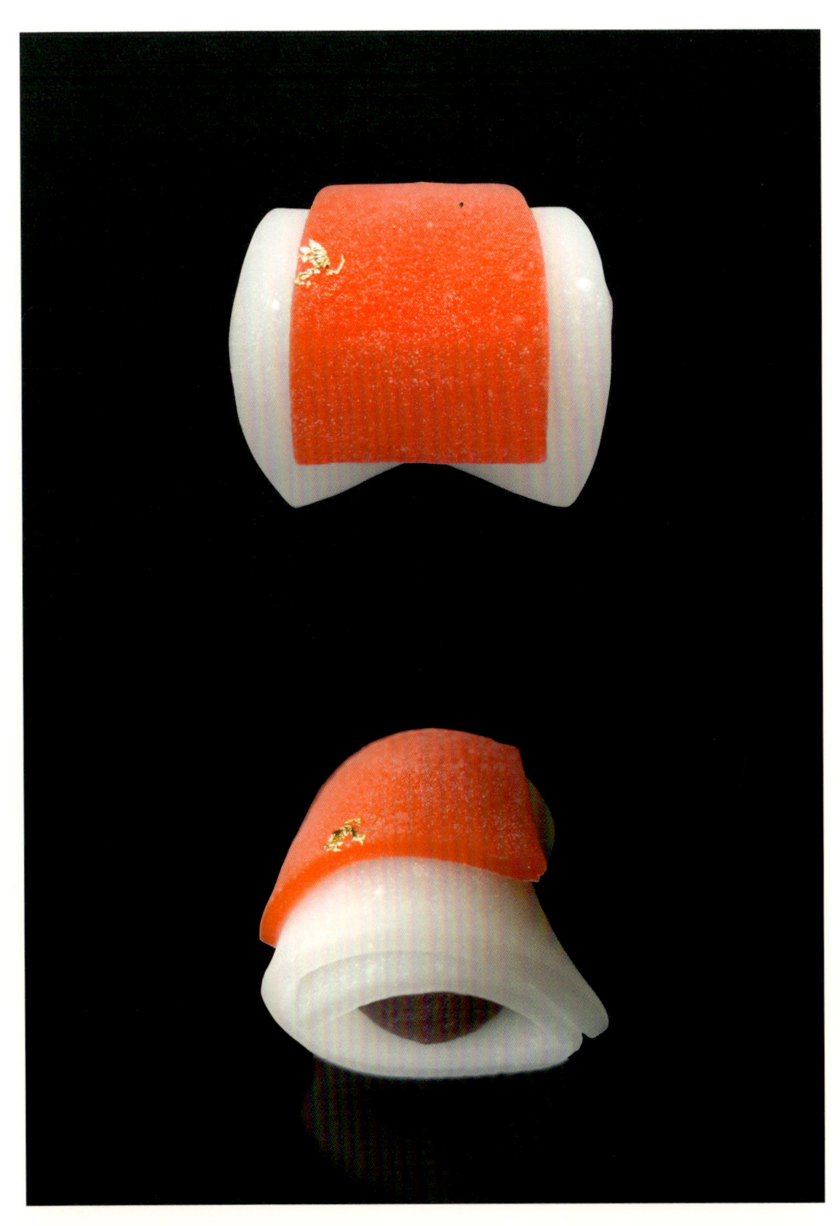

長久堂「十牛の図」 外郎、こしあん
Chokyudo "Jugyu no Zu" *Uiro, Koshian

通し矢 Archery Competition

毎年、成人の日に三十三間堂で行われる「通し矢」で成人を迎える女性の姿はなんとも凛々しい。

Awe-inspiring: A young woman participating in the annual archery competition held at the Sanjusangendo temple on coming-of-age day.

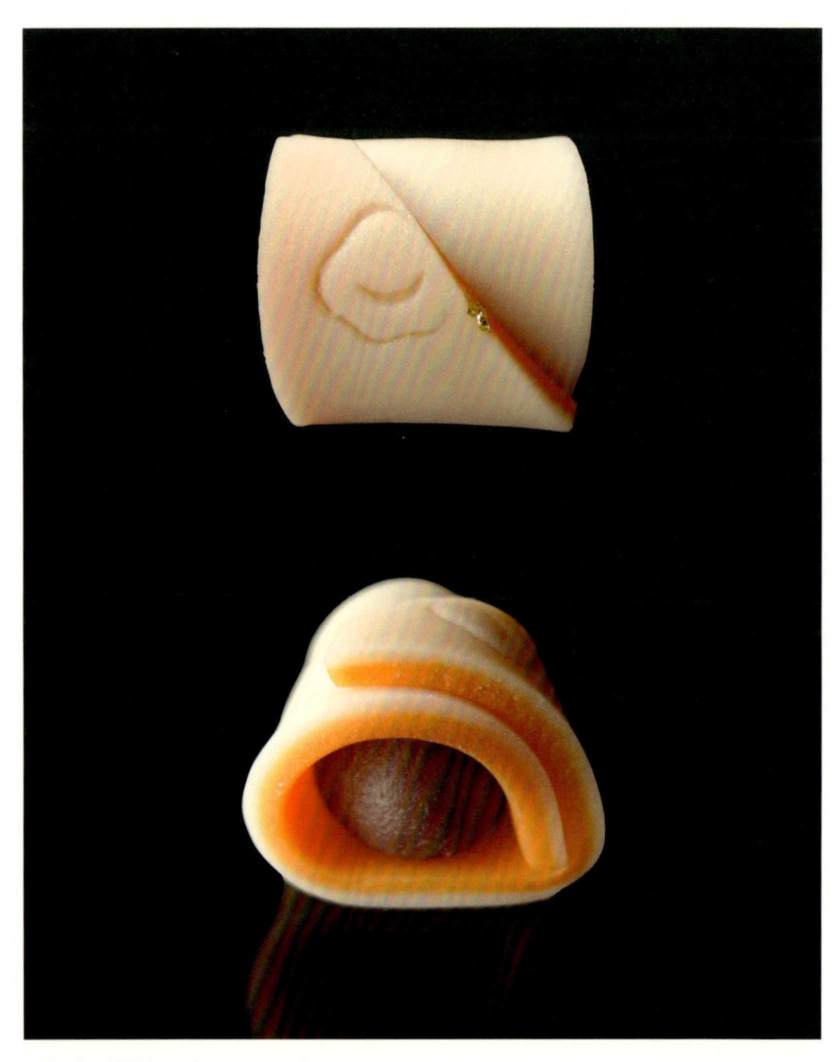

長久堂「袖止め」 こなし、赤こしあん
Chokyudo "Sodedome" *Konashi, Akakoshian

この生菓子の題の袖止めの意味を調べてみました。江戸時代、元服した者が、振袖を縮めて普通の長さにしたことをいい、女性の成人の意にも用いるそうです。

"Sodedome," the name of this moist sweet, refers to the Edo period custom of shortening the flowing sleeves on a young man's kimono when he came of age.

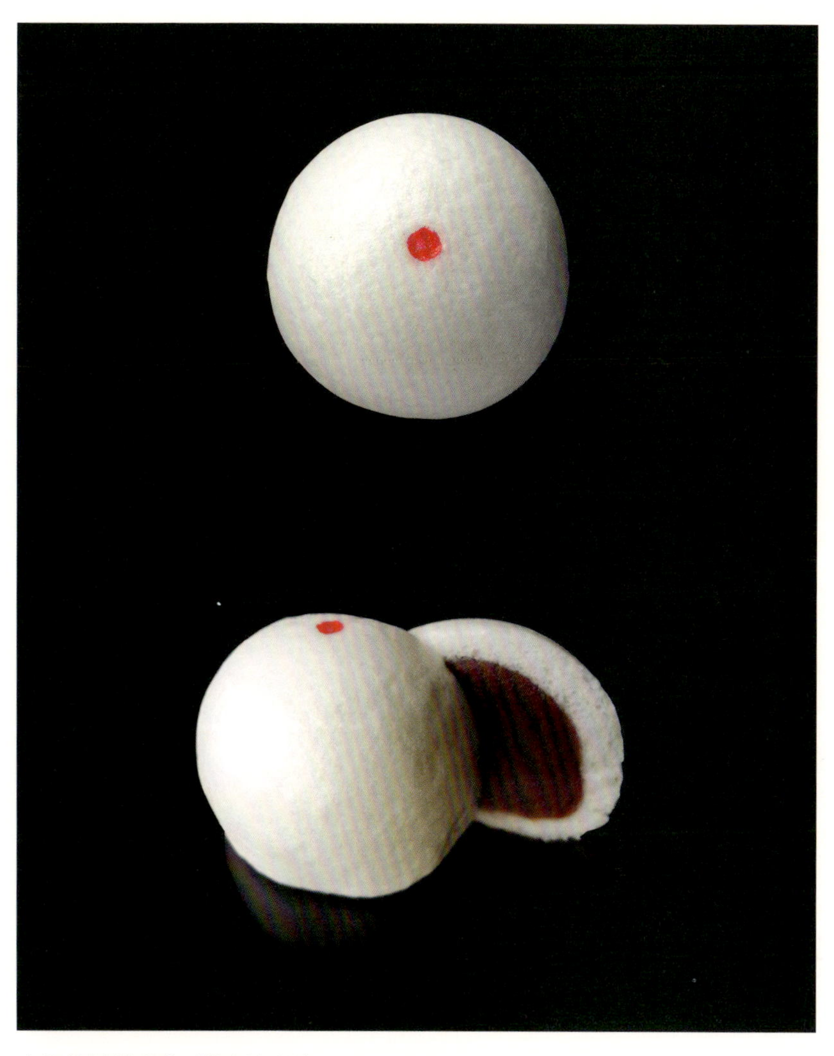

京都鶴屋鶴壽庵「笑顔上用」 上用、こしあん

Kyoto Tsuruya Kakujuan "Hohoemi Joyo" *Joyo, Koshian

この「笑顔上用」というのは、成人式の頃にいただく上用なんだそうです。
年中売っていたらいいのにね。

While "Hohoemi Joyo" is for coming-of-age day, shouldn't it be available throughout the year?

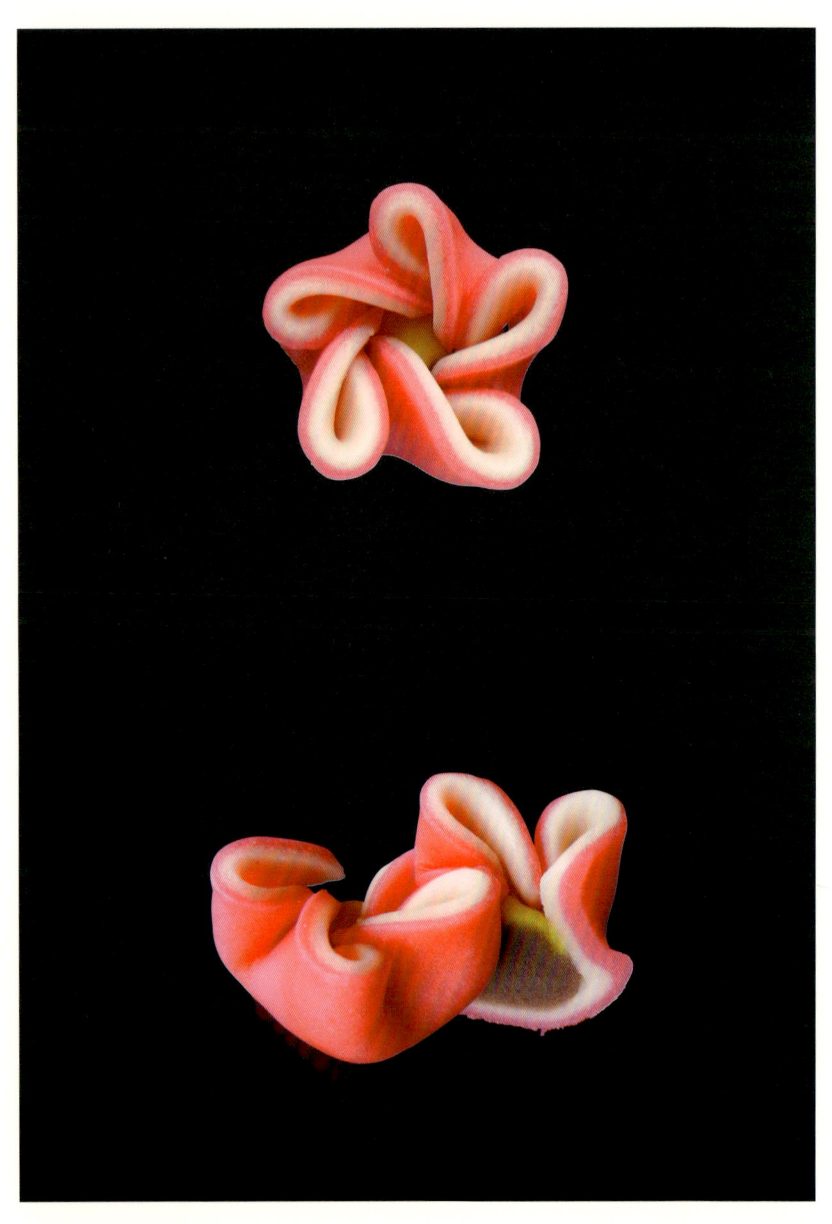

長久堂「初天神」　こなし、こしあん
Chokyudo “Hatsutenjin” *Konashi, Koshian

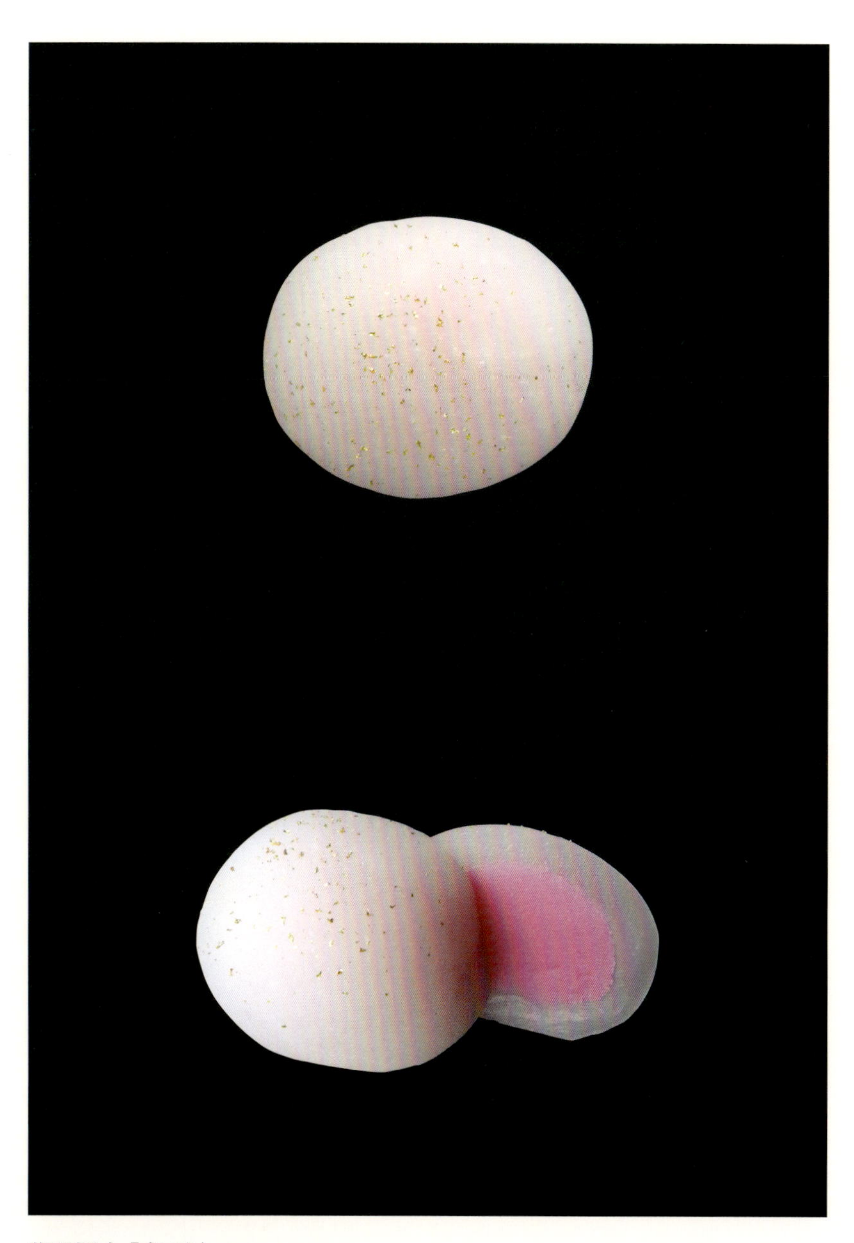

紫野源水「東 雲」 外郎、しろあん
Murasakino Gensui “Shinonome” *Uiro, Shiroan

京都鶴屋鶴壽庵「福寿草」　こなし、黄あん
Kyoto Tsuruya Kakujuan “Fukujuso” *Konashi, Kian

福寿草

春を告げる花の代表。そのため元日草（がんじつそう）や朔日草（ついたちそう）という別名があります。

The golden flower of *Adonis ramosa* flower is the harbinger of the new year. Its names in Japanese include *fukujuso* (“good fortune and long life plant”), *ganjitsuso* (“New Year’s day plant”), and *tsuitachiso* (“first day plant”).

長久堂「千代八千代」 こなし
Chokyudo "Chiyo Yachiyo" *Konashi

この上用菓子の題は「君が代」の歌詞でおなじみです。
This title comes from the national anthem "Kimigayo," to pray forever and ever prosperity

塩芳軒「北野の春」 上用、こしあん
Shioyoshiken "Kitano no Haru" *Joyo, Koshian

シルエットが丑で、紋は北野天満宮ですね。
You may notice the cow in its shape, with the crest of Kitano Tenman-gu Shrine.

二條若狭屋「紅 梅」 煉切り、白こしあん
Nijo Wakasaya “Kobai” *Nerikiri, Shirokoshian

北野天満宮「梅」“Plum” Kitano Tenman-gu Shrine

「東風吹かば　匂ひおこせよ梅の花　あるじなしとて春な忘れそ」
東風は「こち」と読み、春風またはその字のとおり東風という意味で、この歌は拾遺集に収められている菅原道真の代表歌です。

“When the east wind blows, let it carry your fragrance, o plum blossoms. In your master’s absence, forget not the spring” is one of the best known poems by Sugawara no Michizane.

大寒、梅笑う。
京都御苑の梅林に早咲きの梅が、咲いていた。

On the coldest day of the year, the plum trees smile.
On a visit to the plum grove in the Kyoto Gyoen National Park, yes, the early plum trees were in bloom.

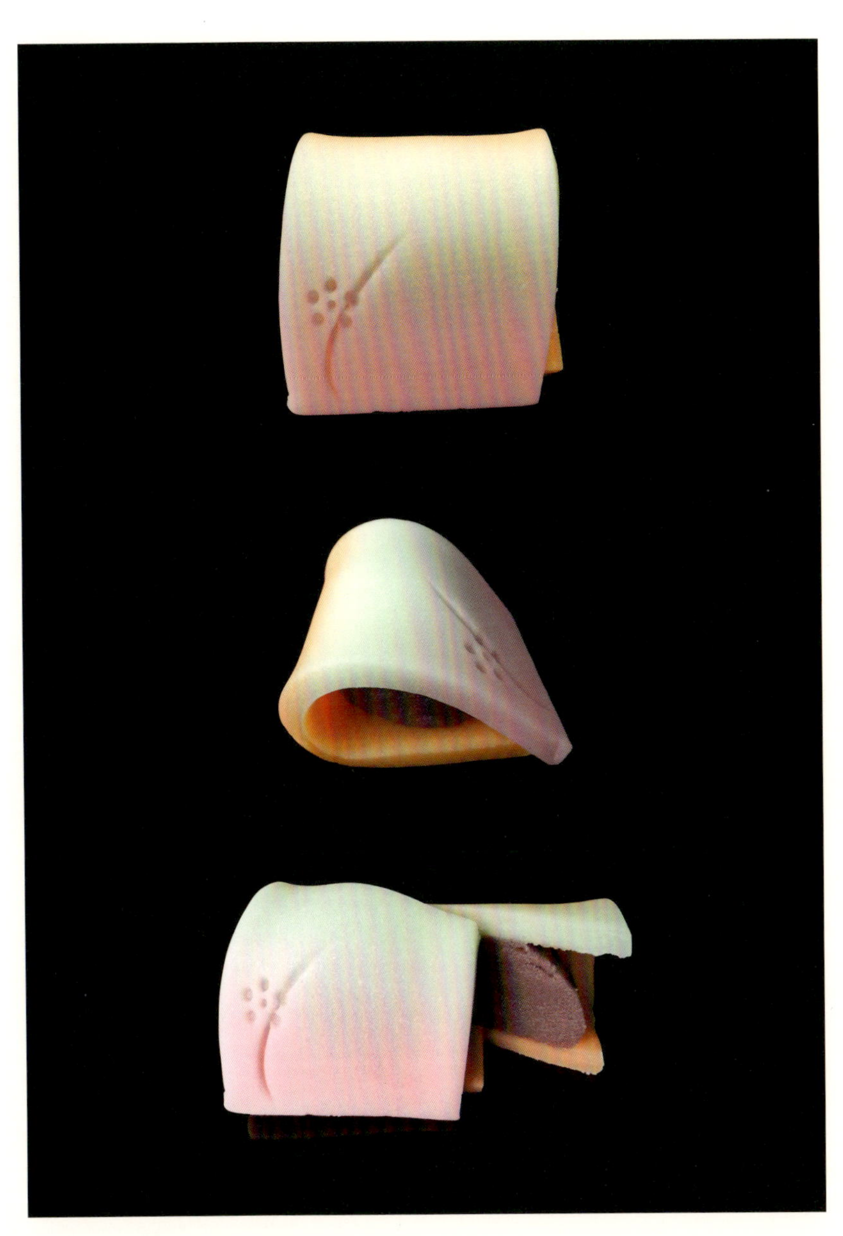

長久堂「一 輪」 こなし(山芋入)、赤ごしあん
Chokyudo "Ichirin" *Konashi with yam, Akagoshian

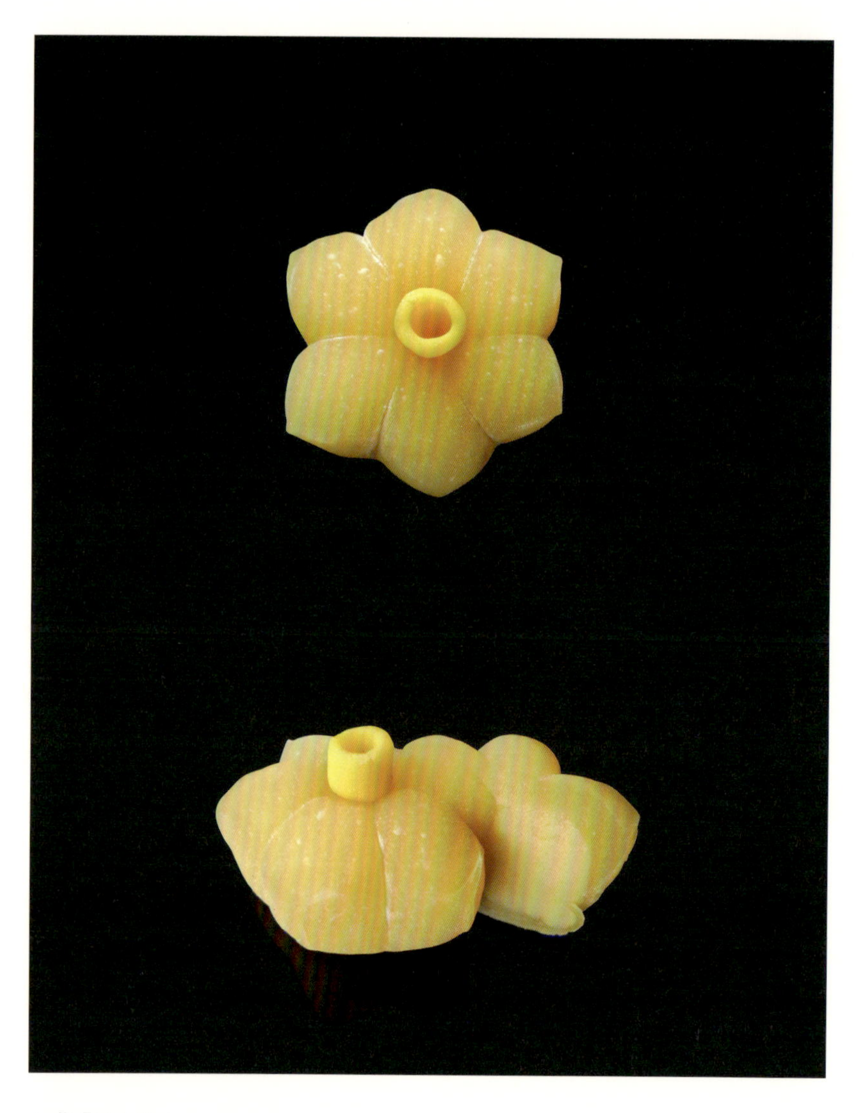

河藤「雪中花」 外郎、黄身あん
Kawafuji "Secchuka" *Uiro, Kimian

雪の中から顔を出し春の訪れを告げる水仙を雪中花と呼ぶそうです。
Daffodil is called Secchuka. One of four beautiful flowers in the snow, daffodils announce the coming of spring.

水 仙　Narcissus

スイセンという名前は、中国での呼び名「水仙」を音読みしたもの。「仙人は、天にあるを天仙、地にあるを地仙、水にあるを水仙」という中国の古典に由来するんだそうです。水辺に咲く姿を見てそう想ったんでしょうね。

Suisen ("water sage"), the Japanese name for narcissus, derives from the Chinese saying, "In heaven, sages are heavenly sages, on earth are terrestrial sages, and in the water are water sages." At the sight of narcissus blooming at the water's edge, I have to agree.

総本家駿河屋「水 仙」 煉切り、白こしあん
Souhonke Surugaya “Suisen” *Nerikiri, Shirokoshian

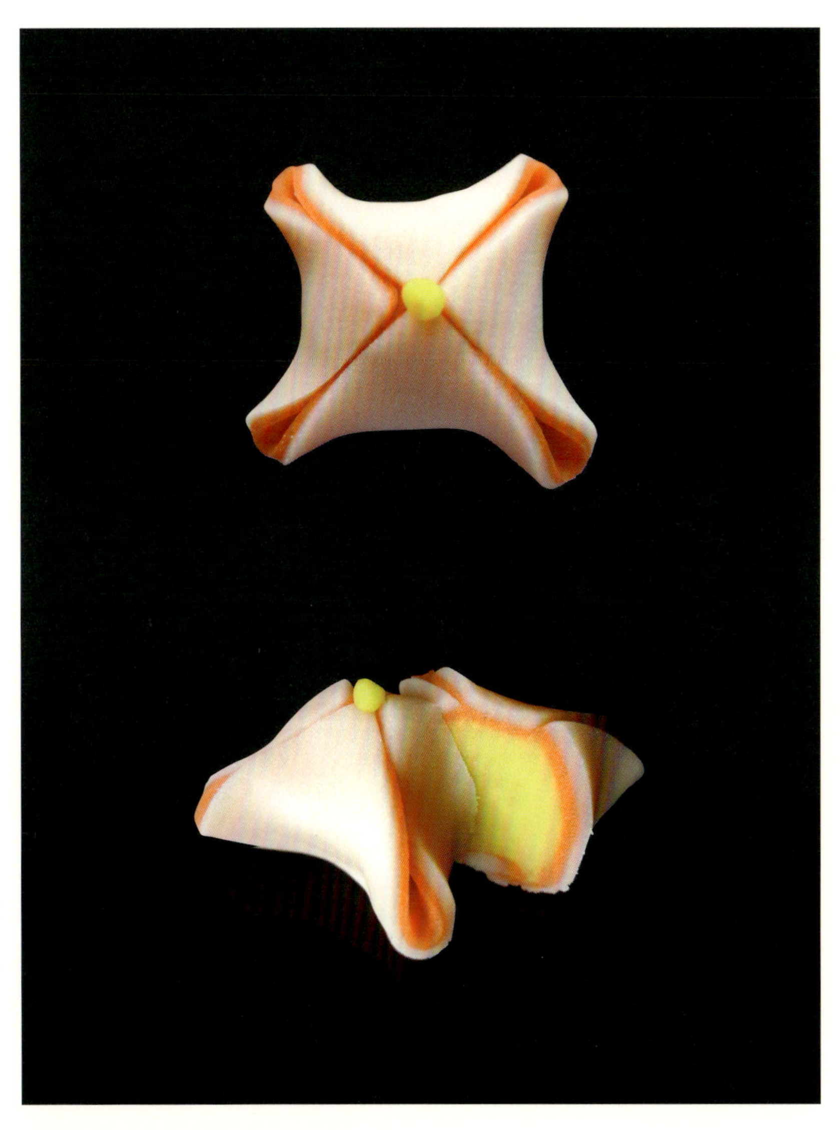

亀屋良長「未開紅」 煉切り、白あん
Kameya Yoshinaga "Mikaiko" *Nerikiri, Shiroan

これから咲こうとしている花を表現しているのか。
This represents the plum buds just before bloom?

寒　椿 Camellia in Winter

年の暮れになると和菓子屋さんやお餅屋さんの店頭では、「椿餅」という貼り紙が目につきます。もうそんな季節なんだ……椿餅は、平安時代からあるそうなんです。道明寺生地の中にこしあんが入り、椿の葉っぱでサンドイッチしたものです。もっとも平安時代には小豆あんなんかはなく、もち米を乾燥させて粗めに挽いたもの（今の道明粉）を水に浸して絞り、甘葛（あまずら）（蔦の汁を煮詰めたもの）を生地に入れたものだそうです。

The sign “*Tsubakimochi*” caught my eye: Is it that time of year already? *Tsubakimochi* dates back to the Heian period: Bean paste inside superb *Domyoji mochi*, sandwiched between camellia leaves. Back then, they added *amazura* syrup to the *mochi*, since they lacked sweet azuki bean paste. (*Domoji* meal is made by drying glutinous rice that has been soaked in water and steamed, then coarsely grinding it.)

中村軒「椿 餅」 道明寺、こしあん
Nakamuraken "Tsubakimochi" *Domyoji, Koshian

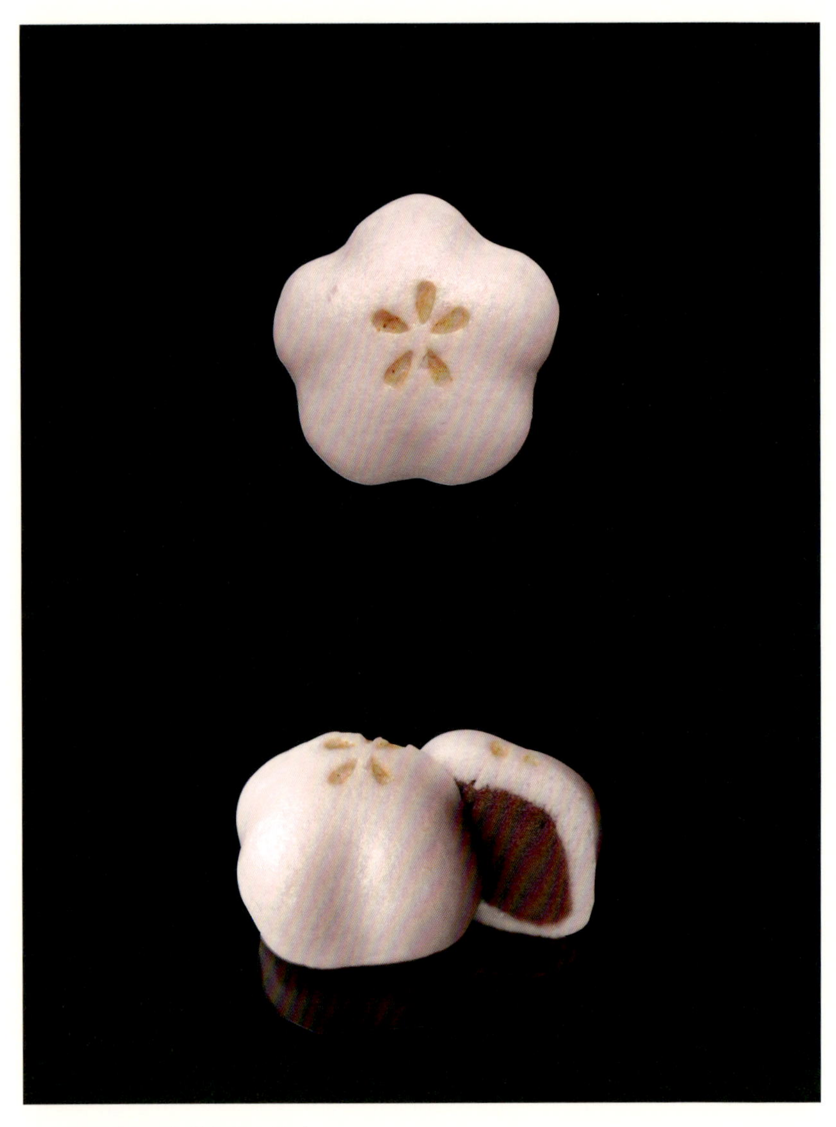

京都鶴屋鶴壽庵「光琳の梅」 上用、こしあん
Kyoto Tsuruya Kakujuan “Korin no Ume” *Joyo, Koshian

北野天満宮の梅の神紋。
The plum blossom crest of Kitano Tenman-gu Shrine.

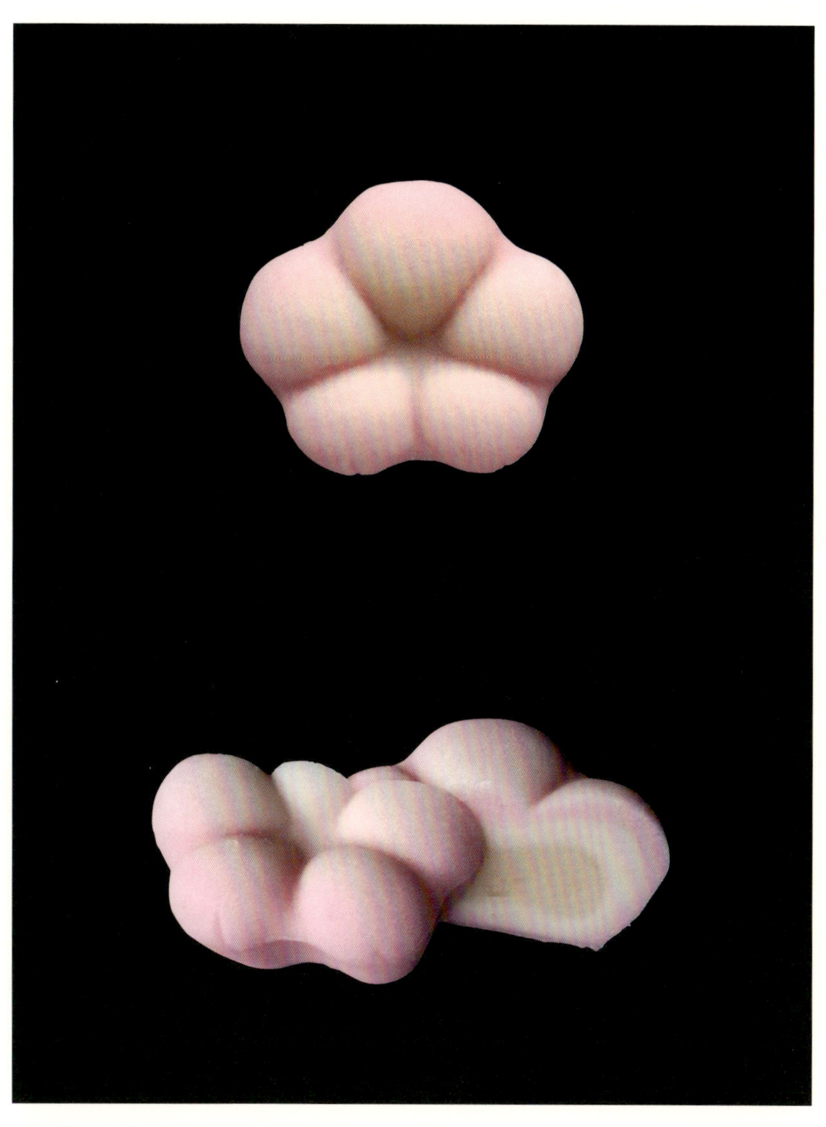

紫野源水「福 梅」 薯蕷煉切り、こしあん
Murasakino Gensui “Fukuume” *Joyo nerikiri, Koshian

総本家駿河屋「鬼は外」　小豆かのこ、粒あん
Souhonke Surugaya "Oni Wa Soto" *Azuki kanoko, Tsubuan

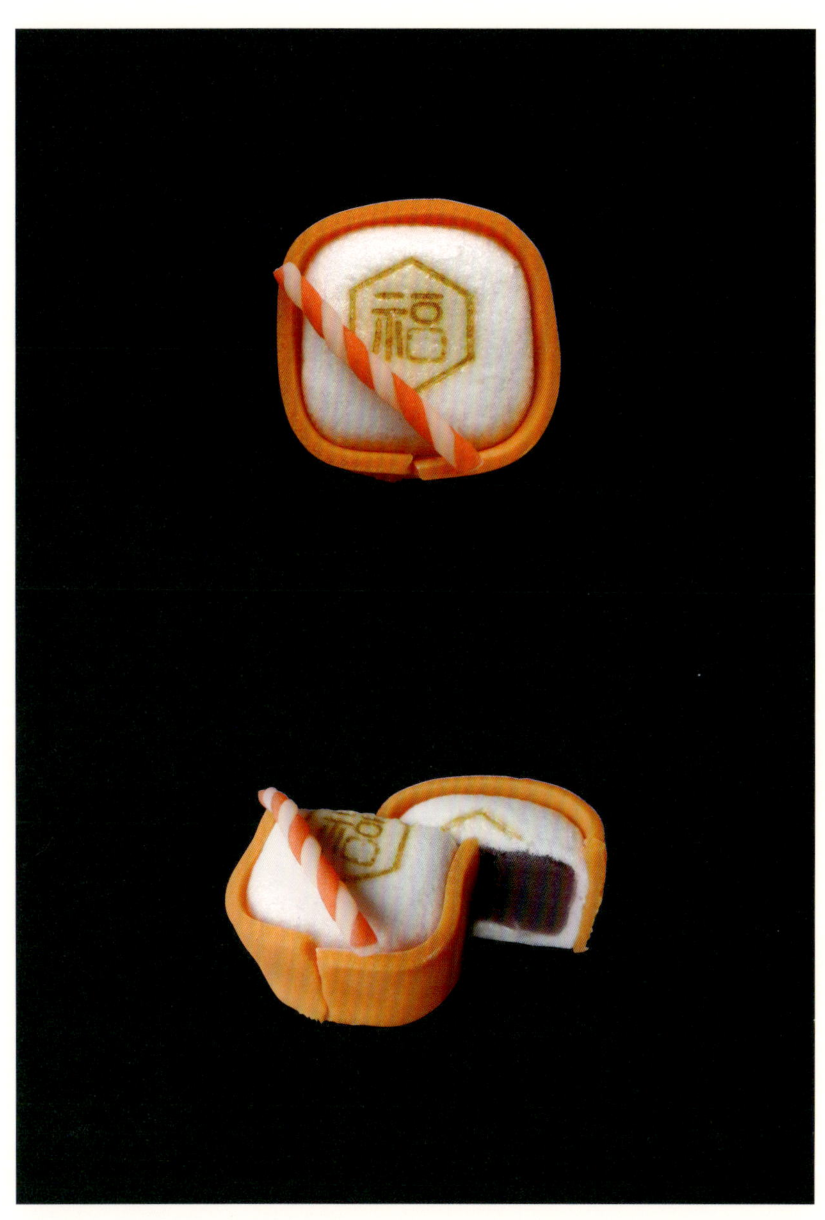

亀屋良長「福ハ内」　上用、こしあん

Kameya Yoshinaga “Fuku Wa Uchi” *Joyo, Koshian

長久堂「福は内」 上用、こしあん
Chokyudo “Fuku Wa Uchi” *Joyo, Koshian

節分は、二十四節気の立春の前の日ということになります。
この日は冬から春に変わる大事な日、その変わり目がとても不安定になるのです。その時に魔物や疫鬼が人間を襲うのでそれを祓うために、追儺式(ついなしき)を行ったそうです。

Setsubun, just before the beginning of spring, is a day when demons and evil spirits may attack people. An exorcism is thus performed, tossing soybeans while chanting, “Good fortune inside, evil spirits outside.” (Hence the beaniness of these seasonal sweets.)

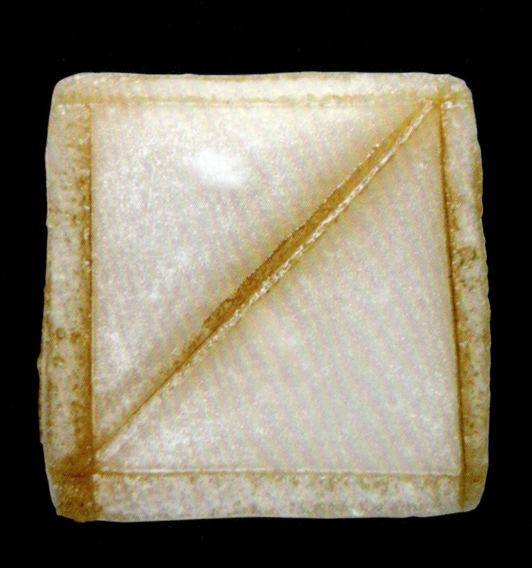

京都鶴屋鶴壽庵「厄払い」 月餅、きなこあん
Kyoto Tsuruya Kakujuan “Yakubarai” *Geppei, Kinakoan

吉田神社の追儺式　Yoshida Shrine “Tsuina shiki”

追儺式は、日が暮れてから本殿前で行われます。もともと平安期の宮中で大晦日に行われていた儀式で、古式に則って行われています。普通の「鬼やらい」と違い、見ていてとても神秘的です。節分の日、季節の変わり目には魔物や疫鬼から家族の身を守らないといけないので対抗策を書いておきます。

1. 豆をまく

これが一番オーソドックスな方法です。豆は魔滅ともいい厄除けパワーがあります。大豆を水に浸けてふやかし、それを乾燥させて煎るとパチパチと音が出ます。魔物はこの音に弱く、でんでん太鼓や弓の弦をはじく音なども嫌いらしいです。昔の人は豆から芽が出るのが不思議だったので神様の力が豆に宿っていると思っていたようです。

2. 鰯を柊の枝に刺して門口に吊るす

節分の時に来る鬼は聞鼻と聞撫いう二匹だそうです。

聞鼻は、鰯のにおいや焼く煙に弱い。聞撫という鬼は、子供が夜おそく手洗いに行くと「聞撫にお尻を撫でられる」といいます。

3. 笑う

魔物や疫鬼は、人間が笑っていると弱るらしい。この方法が一番いいかもしれませんよ。おかしくなくても笑うこと、笑っていると気分が明るくなりますし、免疫力も上がります。

The exorcism, a ritual with roots in the Heian period imperial court, is performed after sunset, in front of the main hall of the shrine. Unlike the usual "chasing out demons" performance, it has an air of mystery. Here are my notes on procedures for protecting one's family from demons and other evil spirits at this turning point in the year:

1. Toss beans

This is the most orthodox procedure. The soy beans (*mame*) acquire their power to exorcise through a pun on "extinguishing demons" (*mame*). Soybeans are soaked in water so that they swell up, then dried and parched, so that they make a popping sound when tossed. Evil creatures are afraid of that noise, as well as that of the pellet drum or strumming on a bowstring. Perhaps people, mystified at how sprouts emerge from beans, assumed divine beings lodged in them.

2. Impale sardines on holly branches and hang them at the entrance to your home.

There are two demons who appear at *setsubun*, Kaguhana and Kainade. Kaguhana is vulnerable to the smells of sardines and cooking smoke. Kainade is said to pat children's bottoms when they go to the toilet late at night.

3. Laugh

Demons and evil spirits are also vulnerable to the sound of human laughter. Laugh, even if nothing is funny, and you will cheer yourself up and even increase your resistance to disease. This may be the best method of all!

総本家駿河屋「赤 鬼」 きんとん、粒あん
Souhonke Surugaya “Akaoni” *Kinton, Tsubuan

おかめ節分のおかめ　“Okame” Okame Setsubun

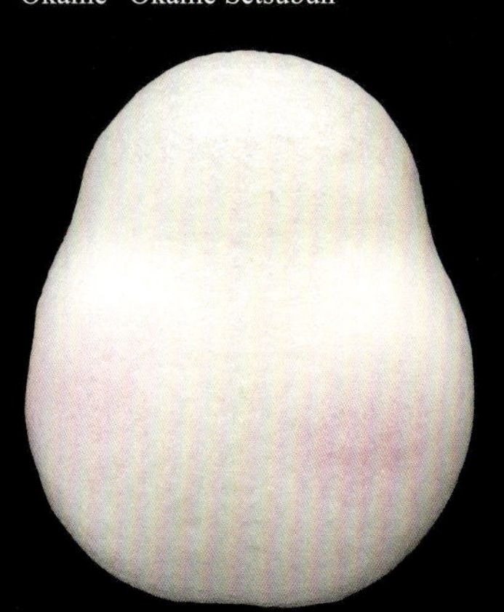

塩芳軒「おかめまんじゅう」　上用、赤ごしあん
Shioyoshiken “Okame manju” *Kinton, Akagoshian

顔を描いていなくても頬がうっすらピンクで、おかめであることが一目瞭然です。
Even without the features drawn in, the pink, plump cheeks tell us at a glance that this is Okame.

白梅の咲く頃　Blooming season of White plum

毎年2月23日は準提堂(じゅんていどう)の「五大力さん」の日(五大力尊法要が行われる日)だ。
Every February 23, the “Five Bodhisattvas of Great Power” are on display at the Junteido.

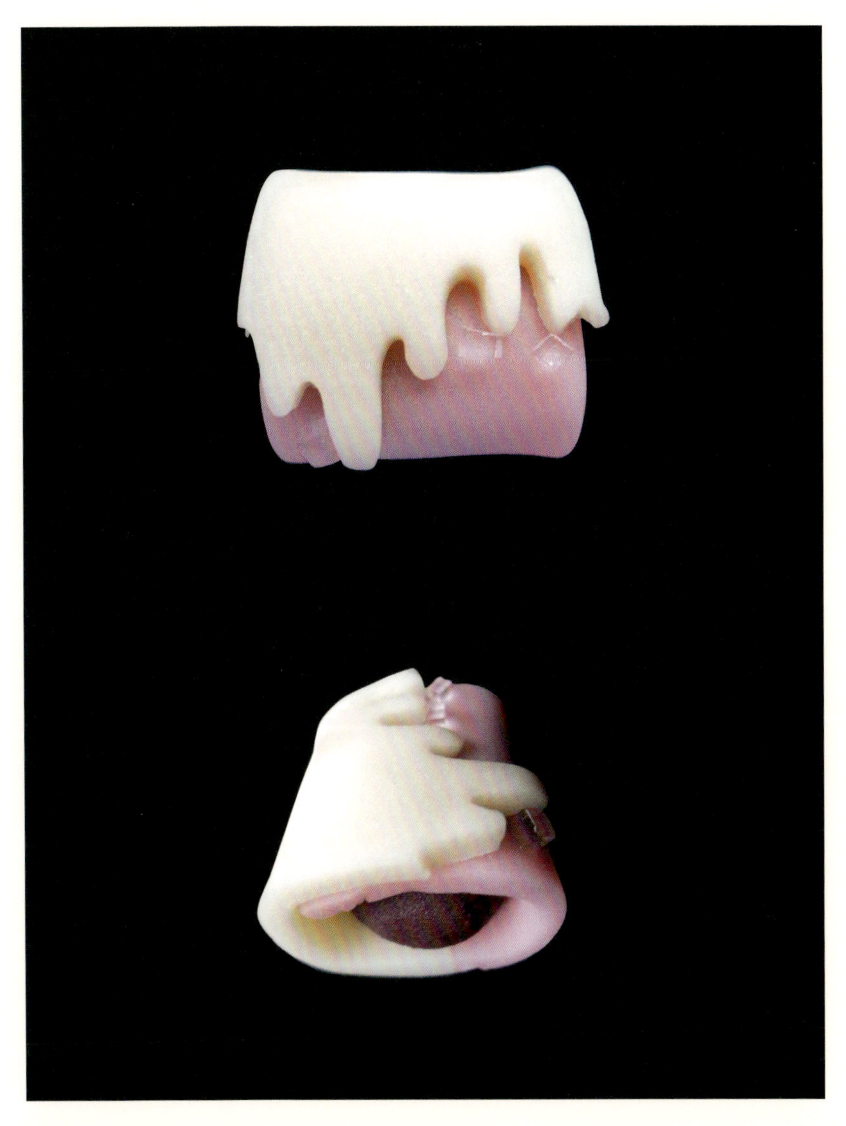

長久堂「春光る」 こなし（山芋、小麦粉）、赤ごしあん
Chokyudo "Haru Hikaru (Spring is shining)" *Konashi with yam and flour, Akagoshian

もう雪は降らへんのかな。
Maybe another snow any more.

清水寺　Kiyomizu temple

つくし Horsetail

河藤「芽生え」 外郎、黄身あん
Kawafuji "Mebae" *Uiro, Kimian

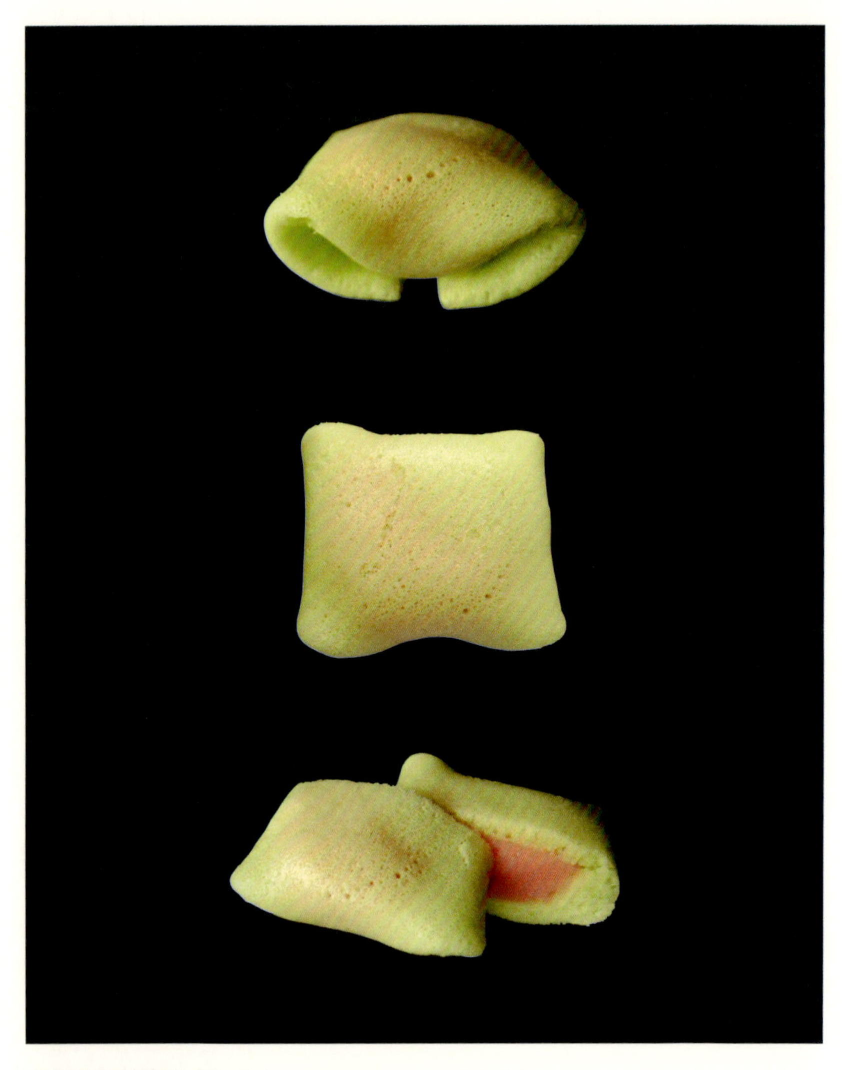

京都鶴屋鶴壽庵「初 音」 青麩焼き、薄紅あん
Kyoto Tsuruya Kakujian “Hatsune” *Aofu yaki, Usubenian

「初音」って、広辞苑でひくとウグイスやホトトギスのその年の最初の鳴き声とあります。
この和菓子、紅餡で梅を、麩焼の色で鶯を表しているそうです。春らしい感じがします。
Hatsune refers to the first warbler or Japanese cuckoo cry heard in spring.
This *wagashi*, represents the warbler by colors of baked Fu and red bean paste, I feel the spring.

紫野源水「春告鳥」　煉切り、白こしあん
Murasakino Gensui "Harutsugedori" *Nerikiri, Shirokoshian

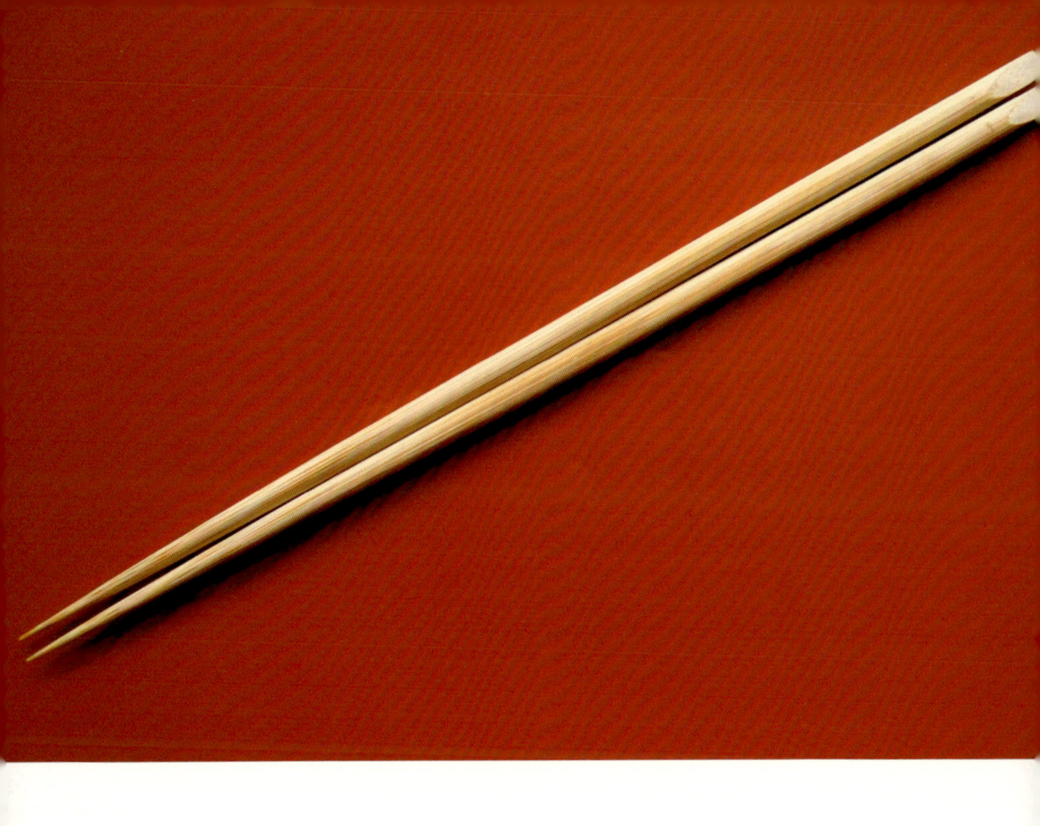

菓子を扱うときに使うお箸、ずっと香道で使う銀の火箸を使っていました。香道には細くていいのですがお菓子ではよく滑り、落としそうになります。
このあいだ、和菓子職人さんとそのことを話していましたら、生菓子のお箸は職人さんが自分で作るものだそうです。それで、作ってもらうことになりました。これでもう滑ったりしないし、細かい部分を修正したりできます。

For working with sweets, I tried the slender silver chopsticks I have long used in the incense ceremony. They are fine for the incense ceremony, but sweets keep slipping off them. Talking with an expert *wagashi tsukasa*, I learned that each makes his own chopsticks, and I had a pair made for myself. True enough, the sweets do not slip down them, and they are designed so that I can make fine adjustments to my work.

長久堂「咲き初む」 こなし、煉切りあん
Chokyudo "Sakisomu" *Konashi, Nerikirian

先日、作ってもらった生菓子用のお箸を使ってみました。とても扱いやすいですよ、なんだかうれしくなります。これで生菓子を落としたりすることはないでしょう。

I tried using chopsticks for Namagashi the other day, I had made. Very easy to deal with, it will be something happy. Will not be dropped or Namagashi with this.

亀屋良長「藪 椿」 山芋入りきんとん、黒粒あん
Kameya Yoshinaga “Yabutsubaki” *Kinton with yam, Kurotsubuan

長久堂「清 雅」 外郎、備中あん
Chokyudo "Seiga" *Uiro, Bicchuan

長久堂「桃花流水」　きんとん、赤ごしあん

Chokyudo “Toka Ryusui” *Kinton, Akagoshian

「桃花流水」というのは、李白の詩中の「桃花流水杳然去」から来ているのでしょうか。

　問余何意棲碧山／余に問ふ何の意ありてか碧山に棲むと

　笑而不答心自閑／笑って答へず　心自づから閑なり

　桃花流水杳然去／桃花流水杳然(ようぜん)として去る

　別有天地非人間／別に天地の人間(じんかん)に非ざる有り

「桃花流水杳然去」の一句は、李白の恬淡(てんたん)とした生きかたが凝縮されている気がします。

Toka ryusui, “peach blossom, flowing water,” is from a verse by the Chinese poet Li Bai that communicates the very essence of his tranquil, aloof way of life.

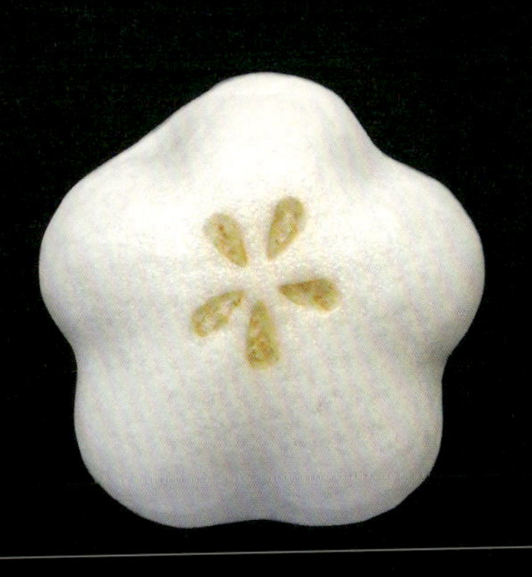

京都鶴屋鶴壽庵「光琳梅」 上用、こしあん
Kyoto Tsuruya Kakujuan “Korinume” *Joyo, Koshian

北野天満宮の蔵、梅のマークがかわいらしい。
Kitano Tenman-gu Shrine warehouse. Cute plum marks.

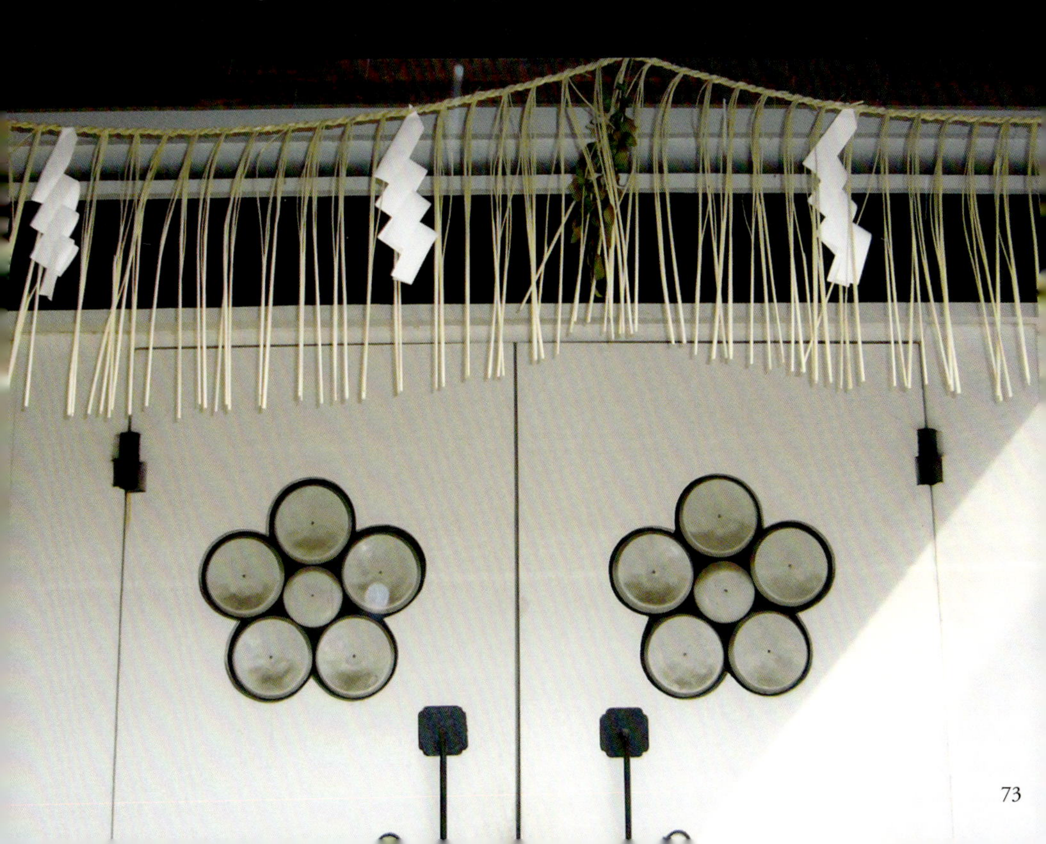

塩芳軒「初参り」 蓬羽二重、黒粒あん
Shioyoshiken “Hatsumairi” *Yomogi habutae, Kurotsubuan

この生菓子の題「初参り」はお宮参りのことですね。
神仏に赤ちゃんが無事に誕生したことを報告し、無病息災をお願いする鎌倉・室町時代からの風習らしい。赤ちゃんの産屋（うぶや）の忌が明ける、男の子は生後三十日目、女の子は三十一日目にお参りするのが一般的といわれています。

“First visit,” the name of this moist sweet, refers to an infant’s first shrine visit. It has been customary for centuries to report a baby’s birth to the buddhas and gods and pray for its health and happiness. That first visit is usually made on the thirtieth day after the birth, for a boy, and the thirty-first for a girl.

紫野源水「薫 風」 煉切り、小豆、こしあん
Murasakino Gensui “Kunfu” *Nerikirian, Azuki, koshian

塩芳軒「寒梅」 上用、こしあん
Shioyoshiken "Kanbai" *Joyo, Koshian

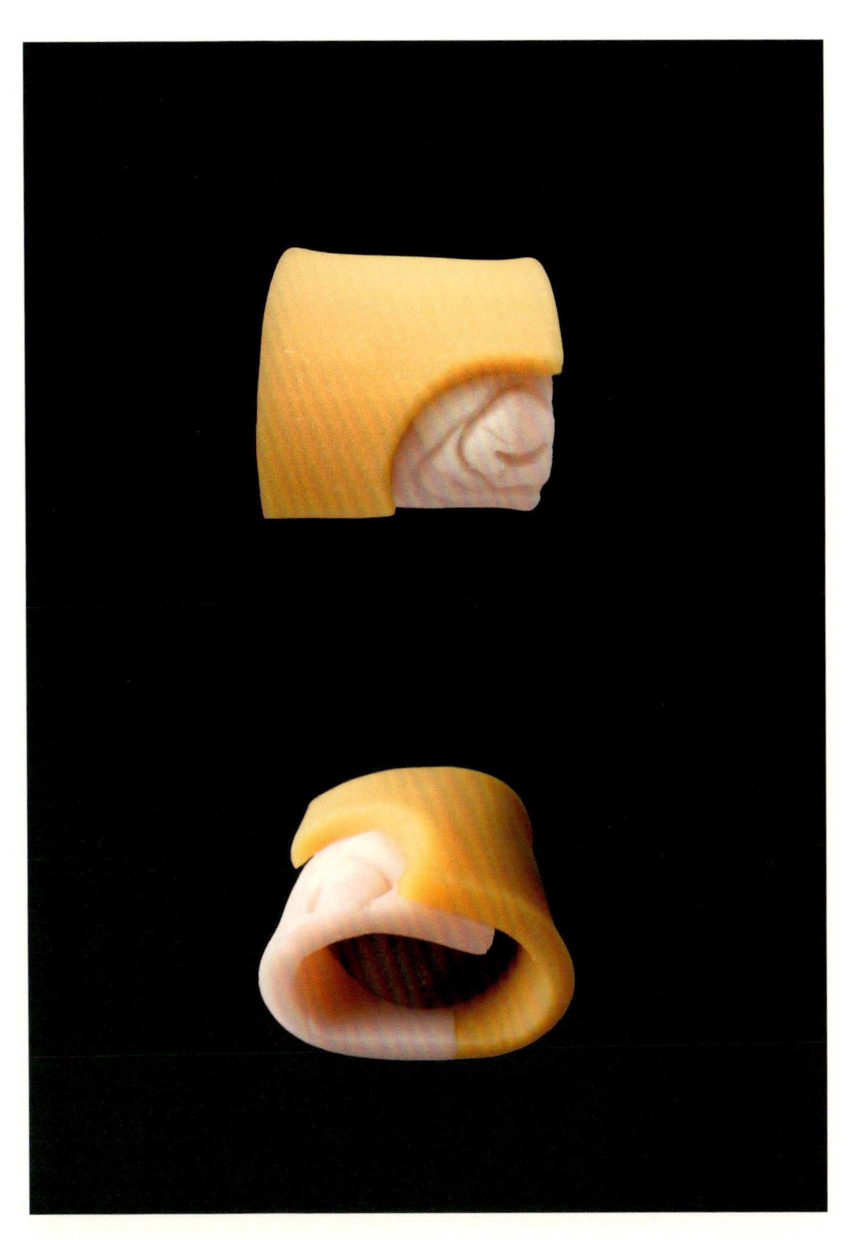

長久堂「窓の梅」　こなし（山芋入）黒こしあん
Chokyudo “Mado no Ume” *Konashi with yam, Kurokoshian

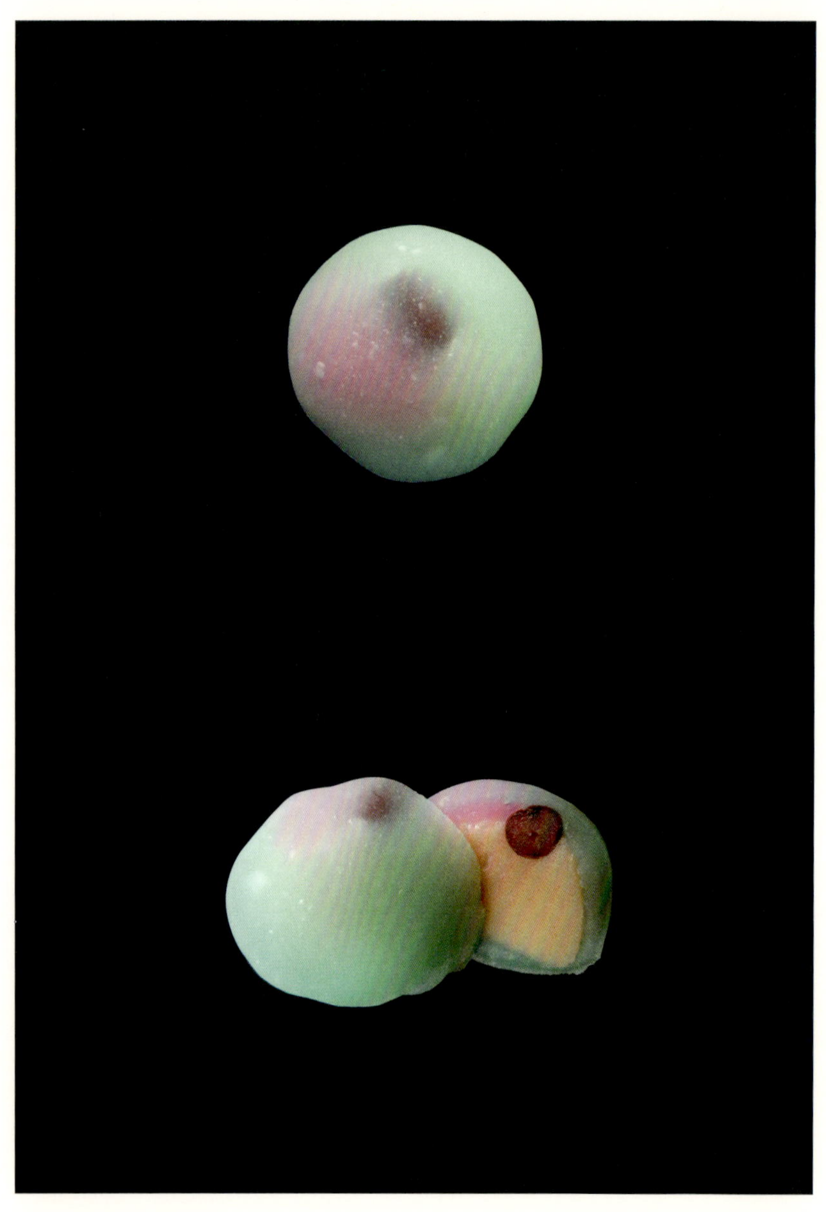

二條若狭屋「水温む」 ういろう、白あん
Nijo Wakasaya “Mizu Nurumu” *Uiro, Shiroan

千本玉壽軒「雪 餅」 つくねきんとん、黄味あん
Senbon Tamajuken “Yukimochi” *Tsukune kinton, Kimian

長久堂「花うさぎ」 上用、こしあん
Chokyudo "Hanausagi" *Joyo, Koshian

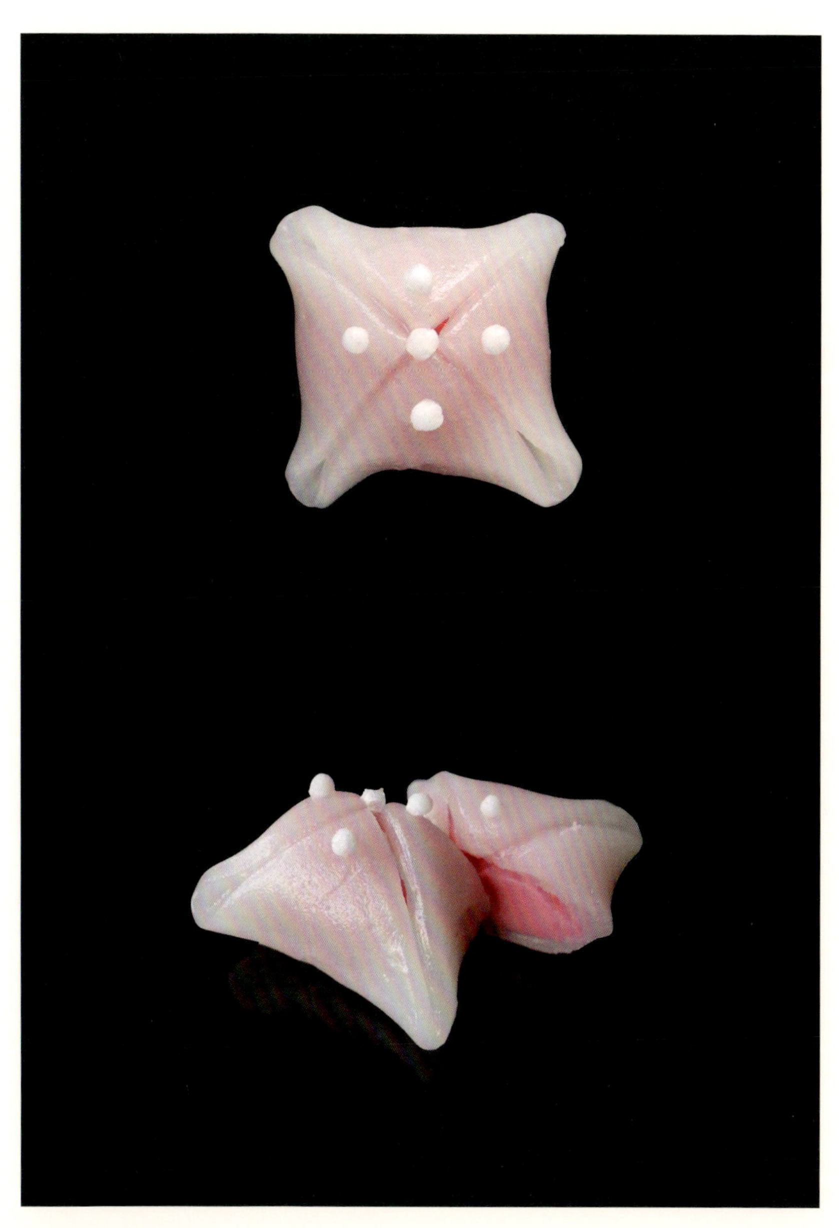

京都鶴屋鶴壽庵「鶯 宿」 外郎、薄紅あん
Kyoto Tsuruya Kakujuan “Oshuku” *Uiro, Usubenian

長久堂「若 草」 外郎（よもぎ、山芋）、黒粒あん
Chokyudo “Wakakusa” *Uiro with yomogi and yam, Kurotsubuan

総本家駿河屋「春の響き」　煉切り、黒こしあん
Souhonke Surugaya "Haru no Hibiki" *Nerikiri, Kurokoshian

紫野源水「ひちぎり二種(白)」 きんとん、粒あん
Murasakino Gensui “Hichigiri (White)” *Kinton, Tsubuan

本来は、餅の中ほどをくぼめて、あんをのせたもの。ひちぎりとは大きな餅を引きちぎること。雛祭には来客が多いので、ゆっくり作業ができず、餅を引きちぎって客に出したことが由来のようです。ひっちぎりともいいます。
Hichigiri is the Kyoto confection served at the Doll Festival.

紫野源水「ひちぎり二種(赤)」　きんとん、粒あん
Murasakino Gensui “Hichigiri (Red)” *Kinton, Tsubuan

上巳（じょうし・じょうみ）の節句は、本来旧暦の３月３日で、
この上巳の日に禊をして、邪気を祓ったことに由来しています。
桃の花が咲く季節であることから、桃の節句ともいわれます。
桃始笑は七十二候の表現で、二十四節気の啓蟄の時候。3月10～14日。
その頃が京都御苑の桃の見ごろでしょうか。

The festival traditionally celebrated on the third day of the third lunar month was originally a purification ceremony to exorcise impurities. Because it coincided with peach trees coming into bloom, it is known as the Peach Festival. The next calendric event is the "awakening of insects," on the sixth of the third lunar month, which is followed by the period known as "peaches begin blooming," from the tenth to the fourteenth of the month. That might be the right time to go see the peach trees in the Kyoto Gyoen National Park in bloom.

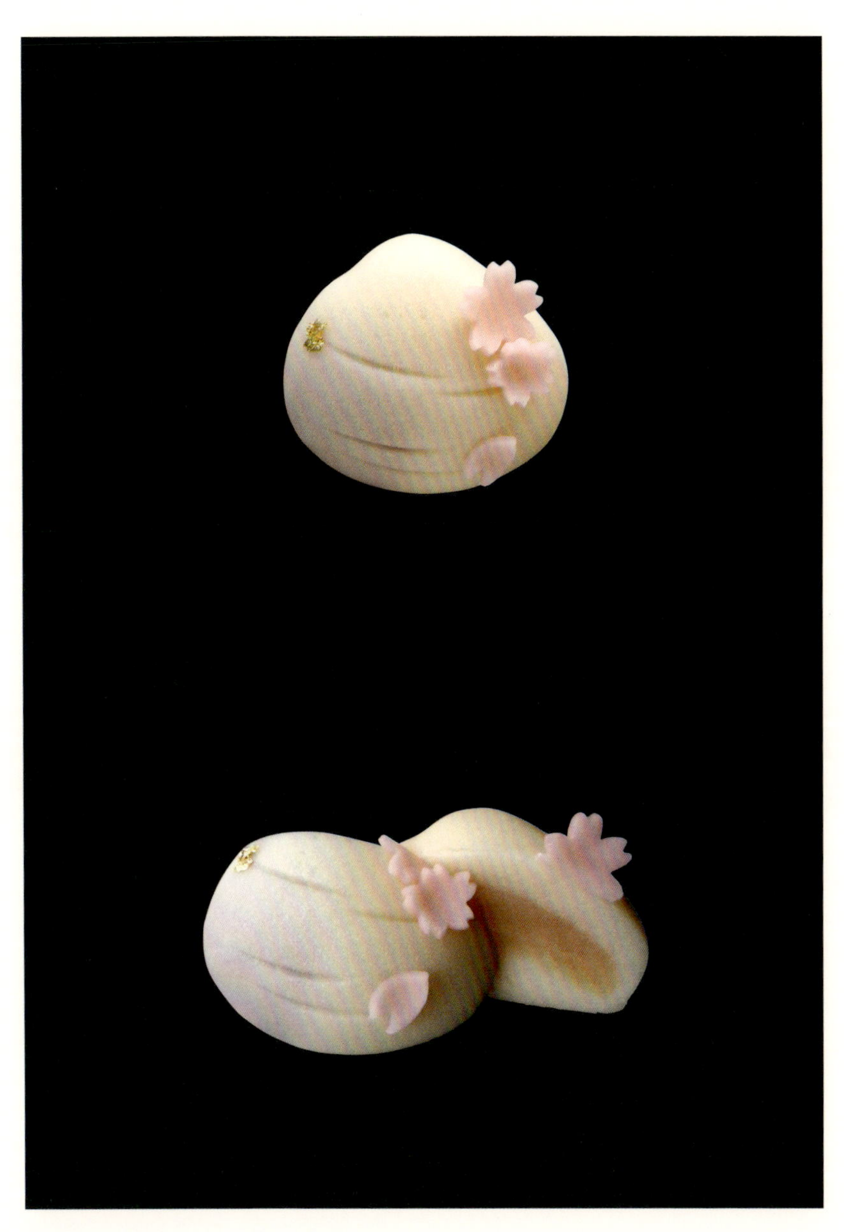

長久堂「貝合わせ」　こなし、備中白こしあん
Chokyudo "Kaiawase" *Konashi, Bicchu shirokoshian

上巳と、宮中で貴族の子供たちが遊んでいた人形遊びとが融合したのが、雛祭りです。江戸時代になると宮中の遊びが江戸にも流行り、次第に女児の祭りに変化していきました。

A purification rite held early in the third lunar month merged with children's playing with dolls in aristocratic households to become a festival held on the third day of the third lunar month. During the Edo period, as the pleasures of the elites spread to the common people, that custom evolved into the Doll Festival, a festival for girls on the same date, with lavish displays of dolls.

紫野源水「引千切」 粒あん、白きんとん
Murasakino Gensui "Hichigiri" *Tsubuan, Shirokinton

京都ではこのお菓子を、おひなさんに必ず供えます。ひちぎりは、ひきちぎりの意味だそうです。おだんごを引きちぎって、ちょうどお玉じゃくしの頭にまるめたあんをのせたような形で。普通はあんまり美味しくないけど、源水製は美味しかった。

In Kyoto, these sweets are always served with the Doll Festival display. The name *hichigiri* is said to be derived from *hikichigiri*, meaning that a dumpling has been shredded up and a dab of bean paste placed in what looks like the bowl of a ladle. These are not usually very tasty, but the Gensui version is delicious.

長久堂「奏」　上用、赤こしあん
Chokyudo “Sou” *Joyo, Akakoshian

紫野源水「早 蕨」 薯蕷、こしあん
Murasakino Gensui “Sawarabi” *Joyo, Koshian

鍵善良房「野の春」 煉切り、こしあん
Kagizen Yoshifusa "No no Haru" *Nerikiri, Koshian

野の春　Spring field

長久堂「衣 手」 外郎、備中あん
Chokyudo "Koromode" *Uiro, Bicchuan

着物の袖、たもとのこと。この語はよく、和歌に用いられています。
Koromode is a term for the sleeve of a kimono often used in poetry.

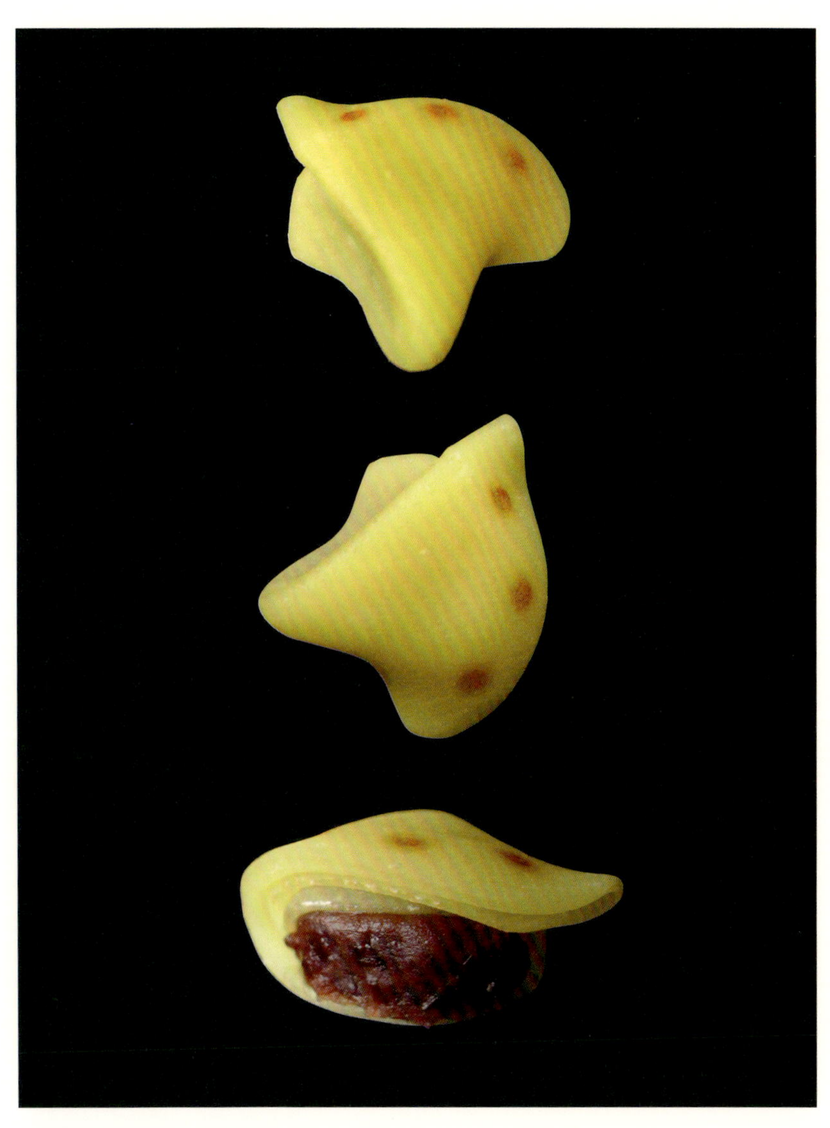

鍵善良房「蝶々」 外郎、粒あん
Kagizen Yoshifusa "Butterfly" *Uiro, Tsubuan

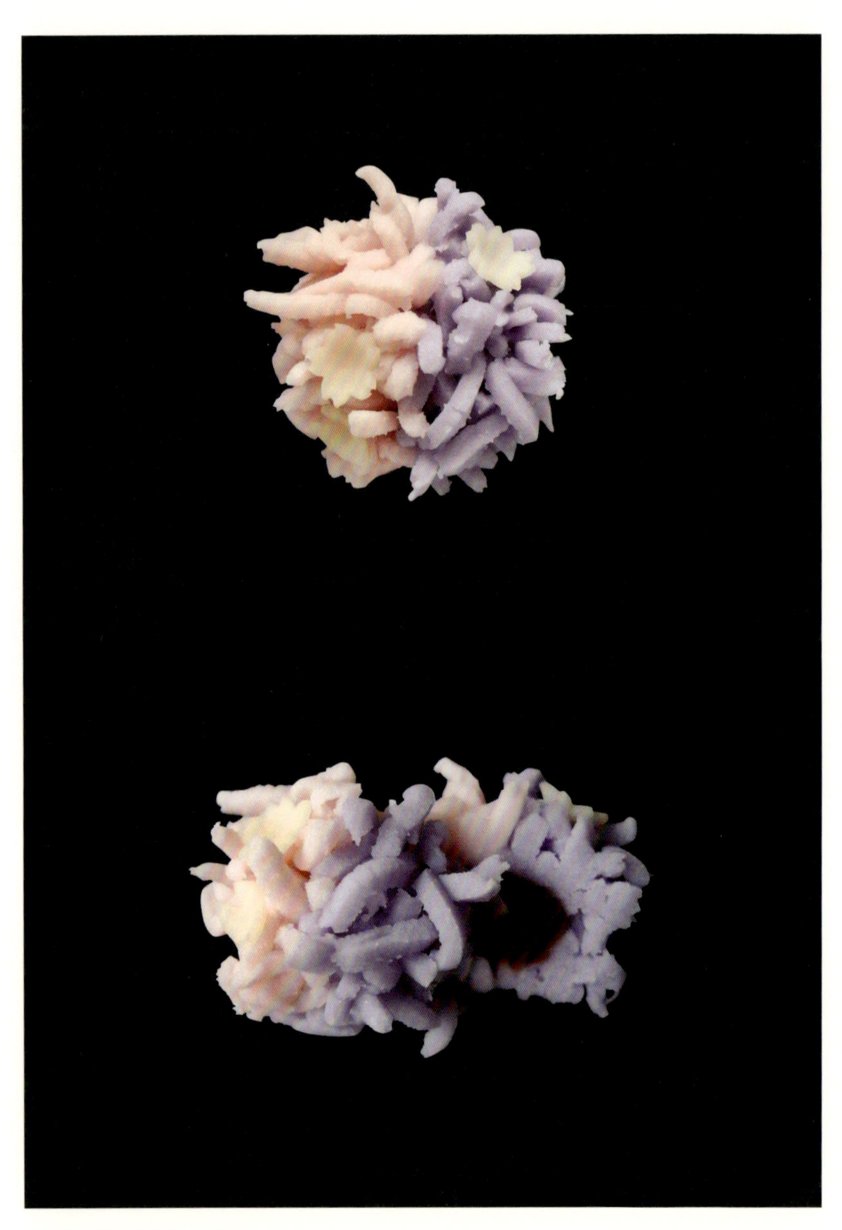

長久堂「春 朧」　煉切製きんとん（山芋入り）、赤こしあん
Chokyudo "Haruoboro" *Nerikiri kinton with yam, Akakoshian

亀屋良長「花 冠」 きんとん、粒あん
Kameya Yoshinaga “Hana Kanmuri(Garland)” *Kinton, Tsubuan

紫野源水「桃の花」 きんとん、粒あん
Murasakino Gensui “Momo no Hana” *Kinton, Tsubuan

ほんとに「桃始笑（ももはじめてさく）」ですね、そろそろ桃の花が見頃かな。
The peach trees must be starting to bloom! We will soon enjoy the sight of their blossoms.

はるの訪れ　Spring comes around

二條若狭屋「ほころび」 煉切り、黒こしあん
Nijo Wakasaya "Hokorobi" *Nerikiri, Kurokoshian

西行の歌で「願わくは花の下にて春死なん　その如月の望月の頃」があります。この歌の頃は旧暦で如月ということは、新暦だと1か月ほど遅れて3月となり、望月（満月）はというと11日頃でしょう。
この西行さん、文治六（1190）年の2月16日（旧暦の）に大阪の南河内の弘川寺で亡くなったのだそうです。前日が涅槃の日、つまりお釈迦さんが亡くなった日なのです。しかもこの日は満月だったということで、偶然にしてはよくできていますね。

The Heian poet, Priest Saigyo wrote, "Let me die under the blossoming trees, in spring, at the full moon in the second month." (He was referring to the lunar calendar; that full moon would occur, in our calendar, in March.)

Saigyo died on the sixteenth of the second lunar month in 1190, at Hirokawa Temple in Minami Kawachi (now Osaka), one day after the anniversary of the death of Shakyamuni, the historical buddha. The moon was full. Well done, Saigyo.

さくらのひとひら　A Petal of cherry blossoms

紫野源水「ひとひら」 煉切り、白こしあん
Murasakino Gensui "Hitohira" *Nerikiri, Shirokoshian

枝垂れ桜　Shidarezakura

桃林から梅林へと丸太町通の堺町御門のほうへ歩いて行くと、枝垂れ桜があります。京都御苑の枝垂れ桜も近衛邸跡の糸桜と同じように、もう咲いていました。満開状態でとてもきれいでした。
京都御苑の北側、今出川御門から入ってすぐのところに近衛邸跡があります。ここの桜は、おそらく京都で一番早い部類の枝垂れ桜ですね。

When you walk from the peach to the plum grove via the Sakaimachi Gomon entrance to the Kyoto Gyoen National Park on Marutamachi Street, you will see this weeping cherry tree. Like the weeping cherries that adorn the former site of the Konoe Palace, it is stunningly beautiful in full bloom.

The Konoe Palace site is on the north side of the park, just inside the Imadegawa Gomon entrance.

京都鶴屋鶴壽庵「御所の春」 きんとん、粒あん

Kyoto Tsuruya Kakujuan "Gosho no Haru" *Joyo, Koshian

三寒四温。昨日は、また寒かった。こうやって春が着実に進行していく。なんだかとても時間の流れるのが早く感じる、密度の濃い毎日。気がつくと夜になっていたりする。この間、鴨川沿いに少しだけ歩いたら、そろそろ小さな雑草の花が色彩豊かに咲いてきていました。もう春ですね。

It was cold again yesterday, but, yes, spring is coming. Somehow every day is so dense with activity that time seems to pass more quickly. Before I know it, night has come again.

Walking along the Kamo River briefly, I found that all sorts of weeds were in flower, most colorfully. Spring is here.

紫野源水「本蕨餅」 小豆こしあん入り
Murasakino Gensui "Honwarabimochi" *Azuki koshian

これはすごく柔らかい、薄い本蕨の食感は芸術的。
こしあんも本蕨をじゃましていなかった。
一見地味なお菓子だけど、食べるのならこういうのがいいですよ。

This is amazingly tender, and the texture of their thin green bracken-flavored coat is an artistic marvel.

The sieved bean paste inside does not clash with the flavor of the bracken. These sweets are rather plain at first glance, but the taste experience they offer is another story.

八重山吹（やえやまぶき）　Japanese kerria

蓮翹（れんぎょう） Forsythia suspensa

街を歩いていると連翹(れんぎょう)がもう咲き始めていました。
黄色やピンク、そしてグリーンは春の感じがしていいですね。黄色といえば、タンポポやミモザ、菜の花もか。もっとも菜の花は、私は色というより味覚のほうで楽しみますが。連翹というのは、(ほとんどが)中国原産なんだそうです。『延喜式』にもレンギョウの名前があるから日本へは、 かなり古くに渡来しているのがわかります。
漢方で「連翹」というと、連翹の実(そんなの見たことないです)を蒸気をとおし天日で乾燥して使うそうです。解熱剤、消炎剤、利尿剤、排膿剤、腫瘍・皮膚病などの鎮痛薬に用います。成分にトリテルペン、モノテルペングリコシド、リグナンを含み、強い抗菌作用があるんだそうです。

The forsythias are already in bloom in the city.
Yellow, pink, green: these are the colors of spring. I associate yellow with dandelions, mimosa, and rape blossoms (though their flavor is even more enjoyable than their color).

Forsythia is said to be native to China, but must have reached Japan fairly long ago. It is mentioned in the *Procedures of the Engi Era*, which were written in the tenth century.

According to *kampo* (traditional Chinese) medicine, forsythia seeds (which I have never seen) are said to be of medicinal use after being steamed, then dried in the sun. They are effective as a febrifuge, an anti-inflammatory, a diuretic, a pus drainage agent, and an analgesic in treating tumors and skin diseases. The seeds contain triterpene, monoterpene glycoid, and lignin, for a powerful antibacterial effect.

真如堂の枝垂れ桜　Weeping Cherries at the Shinnyodo

紫野源水「桜 花」 煉切り、白こしあん
Murasakino Gensui “Cherry Blossom” *Nerikiri, Shirokoshian

真如堂は紅葉の名所ですが、桜もいいんです。枝垂れ桜はもう満開状態でした。私は、山桜と枝垂れ桜の方が八重より好きです。吒枳尼天の桜は、来週が見頃だろうな。ついでに竹中稲荷も見てきたけど、ここも来週かな。今日は風が冷たくてこの間の暖かさがウソのよう。

The Shinnyodo is famous for autumn foliage, but its cherry blossoms are lovely, too — and the weeping cherry trees were in full bloom. I prefer the mountain cherry trees to the varieties with double blossoms. Next week should be about right for going to see the Dekini-ten temple cherries.

I also took a look at the Takenake Inari Shrine; its cherry blossoms will probably open next week. It’s hard to believe it was so warm earlier, given the chilly breeze today.

京都府庁旧本館　Former main building of the Kyoto Prefectural Office.

この旧本館は、明治三十七（1904）年の12月20日に竣工した。昭和四十六年まで京都府庁の本館として実際に使われてきました。創建時の姿をとどめる現役の官公庁建物としては日本最古で、建物の様式は、ルネサンス様式です。なんかしら家に帰ったようななつかしさを感じます。平成十六（2004）年12月10日に重要文化財に指定されています。

This building, which was completed on December 20, 1904, was used as the main building for the Kyoto prefectural government until 1962. Built in a neo-Renaissance style, it is the oldest government building still in use, unaltered, in Japan. It was designated an important cultural properly on December 10, 2004.

I feel a close connection to it: like going home.

醍醐寺のしだれ桜は満開だった！
3月下旬、醍醐寺に行きました。空模様は、目まぐるしく変化し、ベストの空ではなかったのですが、霊宝館のしだれ桜は満開でした。この日はまだ観光客も少なく霊宝館の休憩室にすわってゆっくり眺めることができました。これは本当に見事なしだれ桜ですね。大切に管理されているのがよくわかります。それに応えるよう、ただただ見事に咲いていました。一時暖かい日が続いたので今年は早いのかと思っていたら、また寒が戻り、うまく咲いてくれそうです。

The Daigoji weeping cherry was in full bloom!
I went to Daigoji temple on the twenty-sixth. The weather has been terribly changeable, and the sky was less than ideal, but the weeping cherry by its Reihokan museum was in full bloom. With few tourists out, I was able to sit in the Reihokan rest area and gaze at the tree to my heart's content. What a magnificent tree. I see why they put such care into maintaining it--and it responds by blooming gorgeously. I thought it might have bloomed earlier, since we had a warm spell, but the return of cold weather brought it to bloom perfectly.

灌仏会は、釈迦の誕生を祝う行事で毎年４月８日に行われます。
花祭りというのは、明治につけられた名称なんだそうです。釈迦（ゴータマ・シッタルダ）が旧暦の４月８日に生まれたという伝承に基づいています。
降誕会、仏生会、浴仏会、龍華会、花会式、花祭の別名もあります。

Kanbutsu-e is a ceremony held on the eighth of April to commemorate the birth of the historical buddha, Shakyamuni. The ceremony is celebrated under a variety of names, including, from the Meiji period on, the Flower Festival.

日本では、さまざまな草花で飾った花御堂(はなみどう)を作ってその中に灌仏桶(かんぶつおけ)を置いて、甘茶(あまちゃ)を満たし、その中央に誕生仏の像を安置して、柄杓で像に甘茶をかけて祝います。甘茶は、ユキノシタ科の落葉低木ガクアジサイの変種であるアマチャ、また、その若い葉を蒸して揉んで乾燥させたもの、およびそれを煎じて作った飲料のことです。生薬としては、抗アレルギー作用、歯周病対策等の効果があるそうです。

In Japan, the custom at Kanbutsu-e is to set up a small shrine, decorate it with flowers, and place a basin in it, which is filled with hydrangea tea. Then an image of the Buddha in the form of a child, as he appeared at birth, is placed in the basin and a ladle used to pour the fragrant hydrangea tea over the image.

Hydrangea tea is made from the leaves of *Hydrangea macrophylla var. thunbergii,* by steaming new leaves, crumbling them, and drying them, then making an infusion to drink. It is said to have antiallergenic effects and to prevent gingivitis.

夜桜　Yozakura

亀屋良長「花重ね」　煉切り、白あん
Kameya Yoshinaga “Hana gasane” *Nerikiri, Shiroan

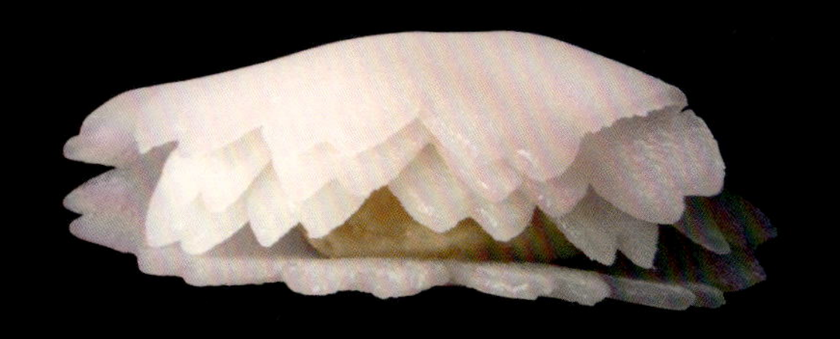

紫野源水「うら桜」 外郎、白小豆粒あん
Murasakino gensui “Urazakura” *Uiro, Shiroazuki tsubuan

平野神社(ひらのじんじゃ)の桜　The Cherry Blossoms of Hirano Shrine

平野神社は、あまりなじみのない神社です。
桜の名所ということしか、知りません。近くの北野天満宮や、わら天神（敷地神社）のほうがなじみがありますが、桜の時期になるとここは人がいっぱい。
神紋が桜というのが面白い。
ここの桜は、平安時代の中ごろ花山天皇によって境内に数千本の桜が植えられたのが起源なんだそうです。もちろん、いまの桜は当時の桜じゃないだろうけど。

Hirano Shrine is surprisingly little known, apart from the beauty of its cherry blossoms. The nearby Kitano Tenman Shrine and Wara Tenjin Shrine attract many visitors, but the Hirano Shrine is busy only in cherry blossom season.

Its crest, interesting enough, is a cherry blossom. The cherry trees originated in several thousand that the Kazan Emperor had planted there in the mid Heian period. Today's trees are, of course, much younger.

清水寺の桜　Cherry blossoms at Kiyomizu temple.

長久堂「花守人」 こなし、赤こしあん
Chokyudo “Hanamoribito” *Konashi, Akakoshian

この上生菓子は、桜を守る人がモチーフなんだろうか、桜の枕みたいです。
This deluxe moist sweets resemble a pillow of cherry wood: are they inspired by the people who protect the cherry trees?

目にしみる緑の葉　Smart Green leaves

カエデの赤ちゃんは、真如堂で見かけました。
このカエデがやがて紅葉の季節を迎え、そして散る。
こうして命が巡っていくから美しいのかも。散るから美しいのです。
I noticed these little maple leaves at the Shinnyodo. In time its leaves will turn red and then fall.
The beauty of the cycle of life is also the beauty of its end.

新緑が見えるベランダからふと見ると木の芽の大木（こんなに大きくなるんだ）の緑がきれい。パワーをもらえるような気がします。

New growth, seen from my veranda.
The green of new shoots on a large Japanese pepper tree is lovely. It seems to bestow power on us.

二條若狭屋「木の芽上用」　上用、こしあん
Nijo Wakasaya “Kinomejoyo” *Joyo, Koshian

これは、不思議な上用です。木の芽がついていますが、木の芽の独特の風味は消えていて
食べても感じられませんでした。

This unusual joyo sweet is decorated with newly budding leaves on it, but their peppery
flavor seems to have vanished.

長久堂「霧 島」 きんとん、粒あん
Chokyudo "Kirishima" *Kinton, Tsubuan

長岡天神の参道の霧島つつじ
Kirishima azaleas line the avenue approaching the Nagaoka Tenjin Shrine.

霧島つつじは、一番早く咲くつつじ。
霧島つつじといえば、この長岡天神のものしか知りませんが、錦水亭にたけのこ料理を食べに行くのがちょうどこのつつじが咲く頃なんです。錦水亭のたけのこの味覚と霧島つつじがわたしには結びついています、焼きたけのこやわかたけ汁が。真っ赤なつつじが情熱的で圧倒的な迫力があります。
樹齢でいうと 100 年以上経っているそうです。

Kirishima azaleas are the earliest to bloom. I only know them from the Nagaoka Tenshin, but the sight reminds me that it's time to enjoy bamboo shoot cuisine at the Kinsuitei. Somehow these azaleas are associated in my mind with the flavor of the Kinsuitei's bamboo shoots.

Their bright red flowers have a passionate, almost overwhelming intensity—and they grow on bushes over a century old.

長岡天神の牡丹　Peony in Nagaoka Tenjin Shrine

京都鶴屋鶴壽庵「牡 丹」 薄紅月餅、白こしあん
Kyoto Tsuruya Kakujuan "Botan" *Benigeppei, Koshian

長岡京の乙訓寺に牡丹を観に行った。
乙訓寺は、牡丹で有名、聖徳太子の時代からあるそうです。いまは、春牡丹の開花時期、牡丹の原産地は中国、元では薬用に栽培されていたそうです。牡丹の根の樹皮部分は「牡丹皮」として、大黄牡丹皮湯、六味地黄丸、八味丸など漢方薬の原料になる。薬効成分はペオノールで消炎・止血・鎮痛などに効くそうです。

I went to see the peonies at the Otokunidera in Nagaokakyo.
That temple has been famous for its peonies since the days of Shotoku Taishi—back in the seventh century! The herbaceous peonies now in bloom are native to China, where they were originally cultivated for medicinal uses. The outer coating of their roots is used in a variety of *kampo* formulations. The active ingredient is paenol, which is said to be an anti-inflammatory, a coagulant, and a pain reliever.

今年も見事に平等院の藤の花が、咲きました。
Beautiful wisteria in a Byodoin temple

鶴屋吉信「藤 宴」 上用、こしあん
Tsuruya Yoshinobu “Toen” *Joyo, Koshian

平等院の藤は、いまが見頃、紫のシャワーを浴びるようです。
藤は、清少納言がとっても好きだったようです。「藤の花、しなひ長く、色よく咲きたる、いとめでたし」とあり、藤がしなやかに長く色美しく咲いているようすを褒めています。

Now is the time to see the wisteria at the Byodoin Temple and revel in their lavender shower.
Sei Shonagon, author of *The Pillow Book*, loved wisteria. Among her list of “splendid things,” she praised “Long flowering branches of beautifully colored wisteria entwined about a pine tree.”

二條若狭屋「藤 浪」 上用、こしあん
Nijo Wakasaya “Fujinami” *Joyo, Koshian

明治四十五（1912）年にできた蹴上の浄水場は、京都市民にとって大切な施設。琵琶湖から疏水を通して京都に水をひき、蹴上の浄水場で処理して各家庭の蛇口に届ける。京都市民は、いつもお世話になっています。前の道路はよく通るのですが、めったに来ることはありません。中に入って、つつじを観るのははじめてですが、毎年５月の連休に一般公開されています。いつからだろう、ここのつつじが注目されるようになったのは。つつじって案外注目されていない、派手なようで地味だし。

Kyoto's Keage water treatment plant, completed in 1912, is a facility of vital importance to the people of Kyoto. Water from Lake Biwa is brought by canal to Kyoto, where it is purified and sent on to the faucets in every home, to our unending benefit. I often go by the site of the plant but almost never venture inside. In fact, this was my first time to enter its precincts and see the azaleas in bloom, when the plant was opened to the public during the early May holidays. When did its vast azalea garden begin to garner attention? Azaleas: splendid, unpretentious, surprisingly overlooked.

千本玉壽軒「蹴上のつつじ」　きんとん、粒あん
Senbon Tamajuken “Keage no Tsutsuji” *Kinton, Tsubuan

二條若狭屋「深山のつつじ」　外郎、白こしあん
Nijo Wakasaya “Shinzan no Tsutsuji” *Uiro, Shirokoshian

亀屋良長「沢の山吹」 外郎、白こしあん
Kameya Yoshinaga “Sawa no Yamabuki” *Uiro, Shirokoshian

新緑の平等院は清々しい！
我々庶民の友である十円硬貨でおなじみの鳳凰堂は千年前、時の権力者関白、藤原道長さんが左大臣、源重信の夫人から譲り受けた別業を、道長の子・頼通が永承七年（1052年）に仏寺に改め、平等院としたそうです。ちなみに別業とは、業が屋敷という意味で、別荘という意味です。

The spring foliage at the Byodoin is so refreshing!
The Phoenix Hall, familiar to us common folk from every ten-yen coin, began as a villa acquired about a thousand years ago by Fujiwara no Michinaga, the Imperial Regent and most powerful political figure of his time, from the wife of the Minister of the Left, Minamoto no Shigenobu. Michinaga's son Yorimichi converted it into a Buddhist temple, the Byodoin, in 1052.

宇治平等院　Uji Byodoin

五月の空　Player for children's health.

長久堂「風薫る」　こなし、赤こしあん
Chokyudo “Kaze Kaoru” *Konashi, Akakoshian

千本玉壽軒「こいのぼり」　こなし、こしあん
Senbon Tamajuken “Koinobori” *Konashi, Koshian

長久堂「もののふ」 外郎、白こしあん
Chokyudo “Mononofu” *Uiro, Shirokoshian

男の子の節句といわれるようになったのは、鎌倉時代以降武士が台頭してからのこと。
粽(ちまき)は、中国戦国時代の楚の愛国詩人屈原の命日に因んだもの。
柏餅に関しては、柏の木は新芽が出るまで古い葉が落ちないということから「家系が絶えない」縁起物として広まったそうです。ここのお店の柏餅は、お餅屋さんやおまん屋さんのと違って、さすがと思う美味しさがある。特にみそあんがいい。年に一度の楽しみです。

The seasonal celebration known as Boys' Day originated sometime in or after the Kamakura period (1185 – 1336), with the rise of the warrior class.

The *chimaki* served on Boys' Day, made of *mochi* wrapped in bamboo leaves, are associated with the patriotic poet Qu Yuan from China's Warring States period.

Kashiwa mochi (oak-leaf *mochi*) are thought to have been popularized as an edible good luck charm: because old oak leaves stay on the tree until the new ones start to emerge, they symbolize an unbroken family lineage. The *kashiwa mochi* prepared by this shop are far different from what run-of-the-mill confectionary shops turn out. They are utterly delicious, and the filling of white bean paste and miso is superb. They are an annual delight.

長久堂「柏 餅」(中) こしあん　(下) みそあん
Chokyudo "Kashiwamochi" *Koshian / Misoan

大田神社の杜若（かきつばた）　Once again it is time to enjoy the rabbit-ear iris at the Ota Shrine.

今年も大田神社の杜若が見頃になってきた。
大田神社は、上賀茂神社の境外摂社（けいがいせっしゃ）です。摂社は上賀茂神社の祭神と縁故の深い神を祀った神社のこと。境内の大田の沢は平安時代から野生の杜若が有名で、国の天然記念物になっています。まさに尾形光琳の世界。
「神山や　大田の沢の　かきつばた　深きたのみは　色にみゆらん」（藤原俊成の歌）
沢ということは、昔は広くて流れがあったんだろうな。

The Ota Shrine, an auxiliary shrine of the Kamigamo Shrine, enshrines a deity closely related to the one enshrined at Kamigamo. Ota marsh, inside the shrine precincts, has been famous since the Heian period for its rabbit-ear iris. It has now been designated a national natural preservation site. Here we are truly in the world of Ogata Korin. “Sacred mountains and the irises of Ota marsh—the depth of our prayers can be seen in their color” — Fujiwara no Toshinari.

The marsh was apparently broader, with a stream flowing through it, in the past.

長久堂「大田の沢」　外郎、白こしあん
Chokyudo "Ota no Sawa" *Uiro, Shirokoshian

梅宮大社の黄菖蒲　This is the Umemiya Grand Shrine

もう少しすると花菖蒲の時期なんだろうな、というシーズンオフの日に訪れましたがもう咲いていました。まだ少し杜若も残っていました。

The Japanese iris will soon be in bloom. When I visited, a few of the rabbit-ear iris were still flowering.

京都鶴屋鶴壽庵「花あやめ」　こなし、黄あん
Kyoto Tsuruya Kakujuan “Hanaayame” *Konashi, Kian

二條若狭屋「花菖蒲」 煉切り、こしあん
Nijo Wakasaya “Hanashobu” *Nerikiri, Koshian

梅雨入りももうそろそろ、花菖蒲はいまが見頃です。
本当は雨に濡れた花菖蒲や睡蓮がいいのですが、傘をさして撮影はちょっと。
観光的にはオフシーズンなので静かに神苑を楽しめます。いまぐらいに見る紫色はとてもきれいに感じます。

— and for the rainy season is finally here.: time to enjoy Japanese iris.
Irises and lotus flowers in the rain are a wonderful sight, but shooting while holding an ummbrella is not fun.

It is a delight to be able enjoy the Heian Shrine garden quietly, since this is an off season for tourism, despite the bad weather. The purples are especially beautiful about now.

いまが見頃の宇治の三室戸寺(みむろとじ)のつつじ
The azaleas at the temple of Mimurotoji, in Uji, are in bloom.

西国観音霊場第十番札所で本山修験宗の別格本山です。宝亀元(770)年、光仁天皇の勅願により、三室戸寺の奥、岩淵より出現された千手観音菩薩を御本尊として創建されました。1200年以上も前のことです。京都では、「花の寺」として知られていて、いまがつつじ。6月にはアジサイ、7月には蓮、11月には紅葉が有名。ここのアジサイは二万株ぐらいあり、満開になるとアジサイの大海原のような状態になります。これはすごいのひとこと。

Mimurotoji is the tenth temple in the Saigoku Kannon pilgrimage route. Built at the order of the Konin Emperor in 770, its main focus of devotion is a Thousand Armed Kannon that emerged from the crags behind the temple, about 1200 years ago. Known as the "temple of flowers" in Kyoto, at this season, it has azaleas in bloom. They are followed by hydrangeas in June, lotus flowers in July, and autumn foliage in November. The temple's twenty thousand hydrangea plants turn it into a sea of blue when they are in bloom—an amazing sight.

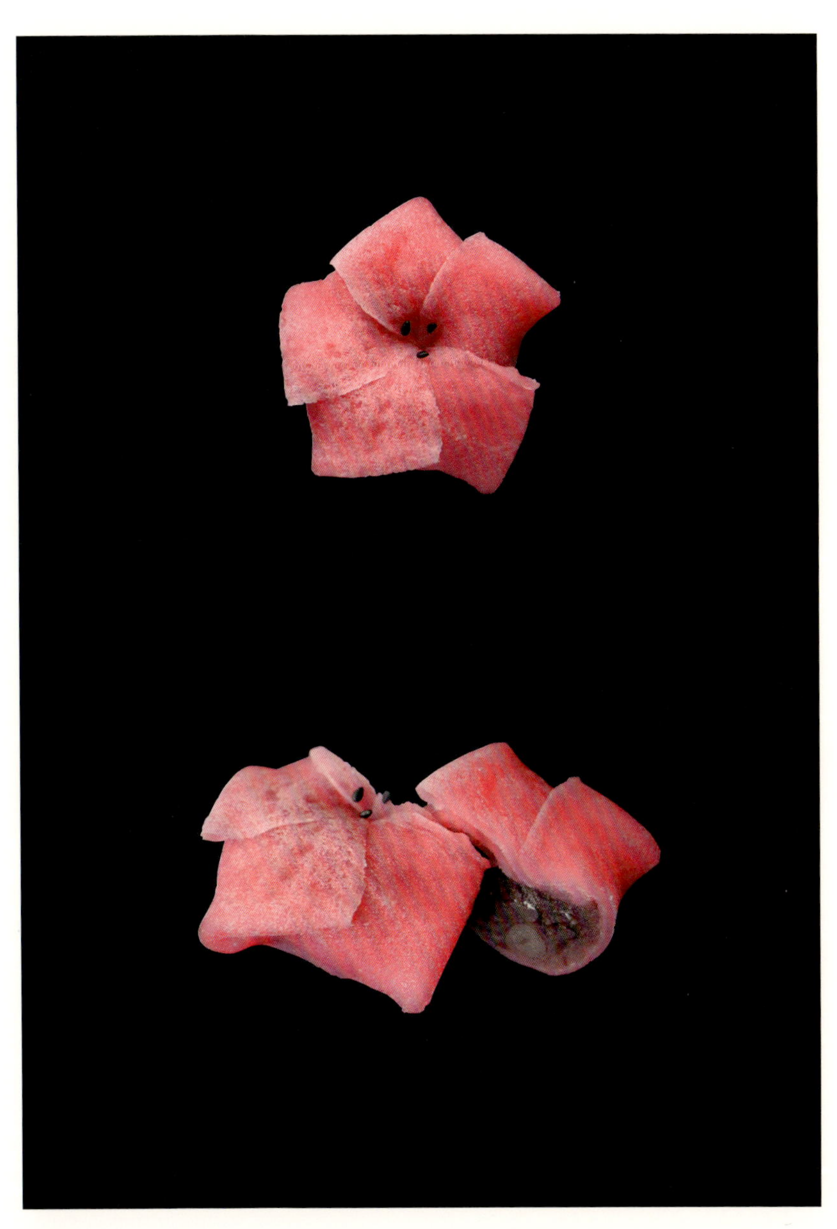

鶴屋吉信「さつき花」 焼皮、つぶあん
Tsuruya Yoshinobu “Satsukibana” *Yakigawa, Tsubuan

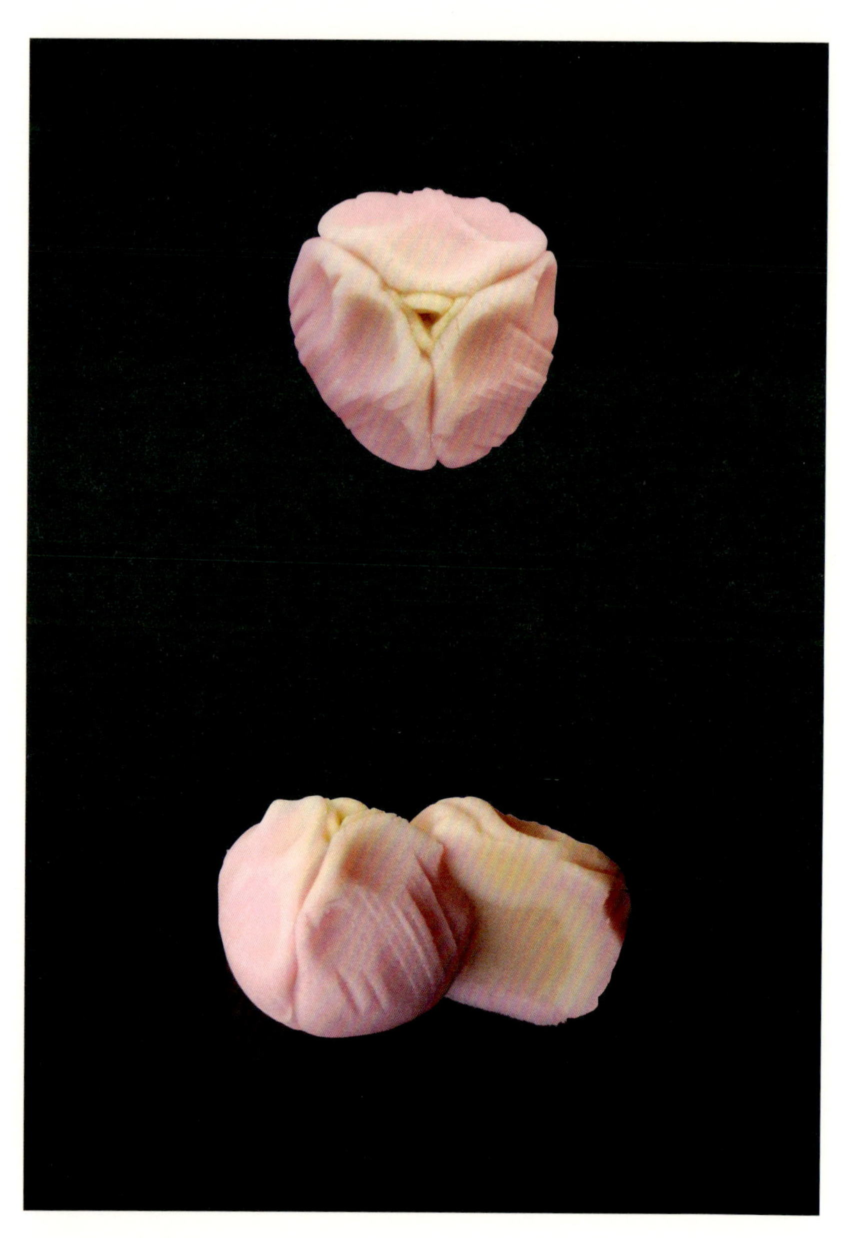

紫野源水「富貴草」 煉切り、白小豆こしあん
Murasakino Gensui "Fukiso" *Nerikiri, Shiroazuki koshian

花水木(はなみずき)の開花前　Flowering dogwood, about to bloom

変わった形をしているな、開いても不思議。
別名、アメリカヤマボウシ。アメリカから来た花なんだそうです。

What odd buds-and unusual flowers.
The flowering dogwood is said to be native to North America.

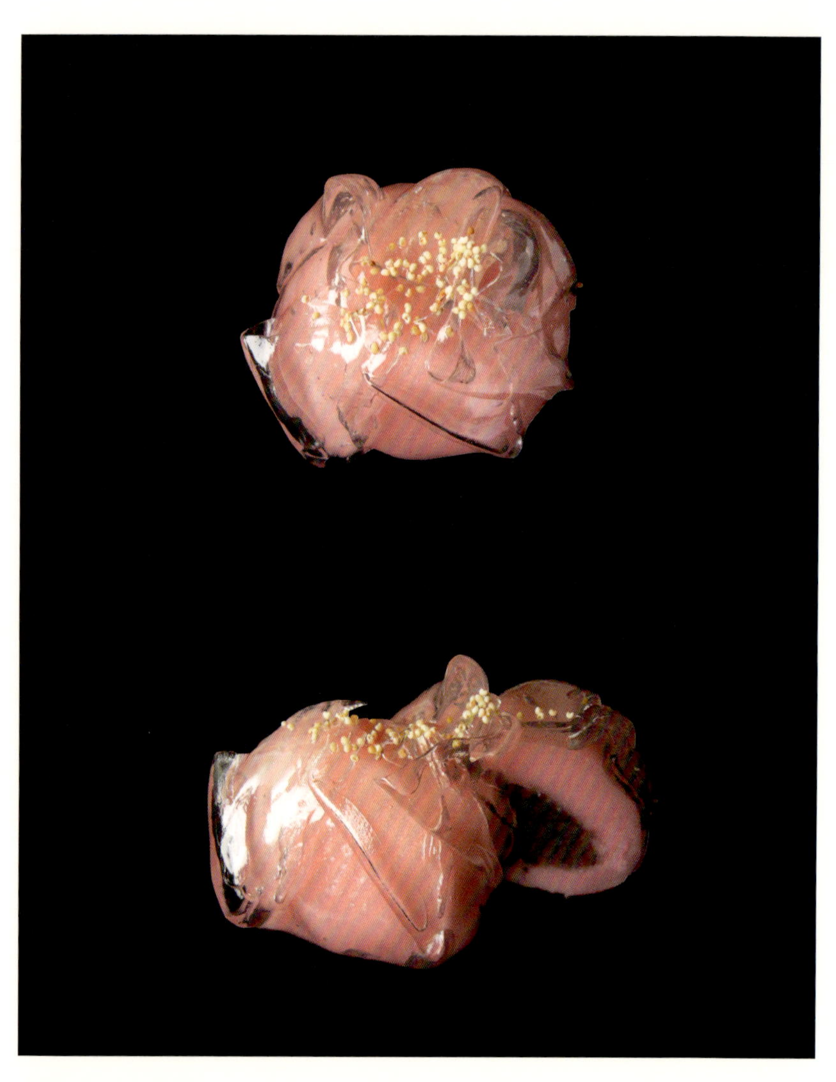

亀屋良長「水牡丹」 寒天、煉切り、つぶあん
Kameya Yoshinaga “Mizubotan” *Kanten, Nerikiri, Tsubuan

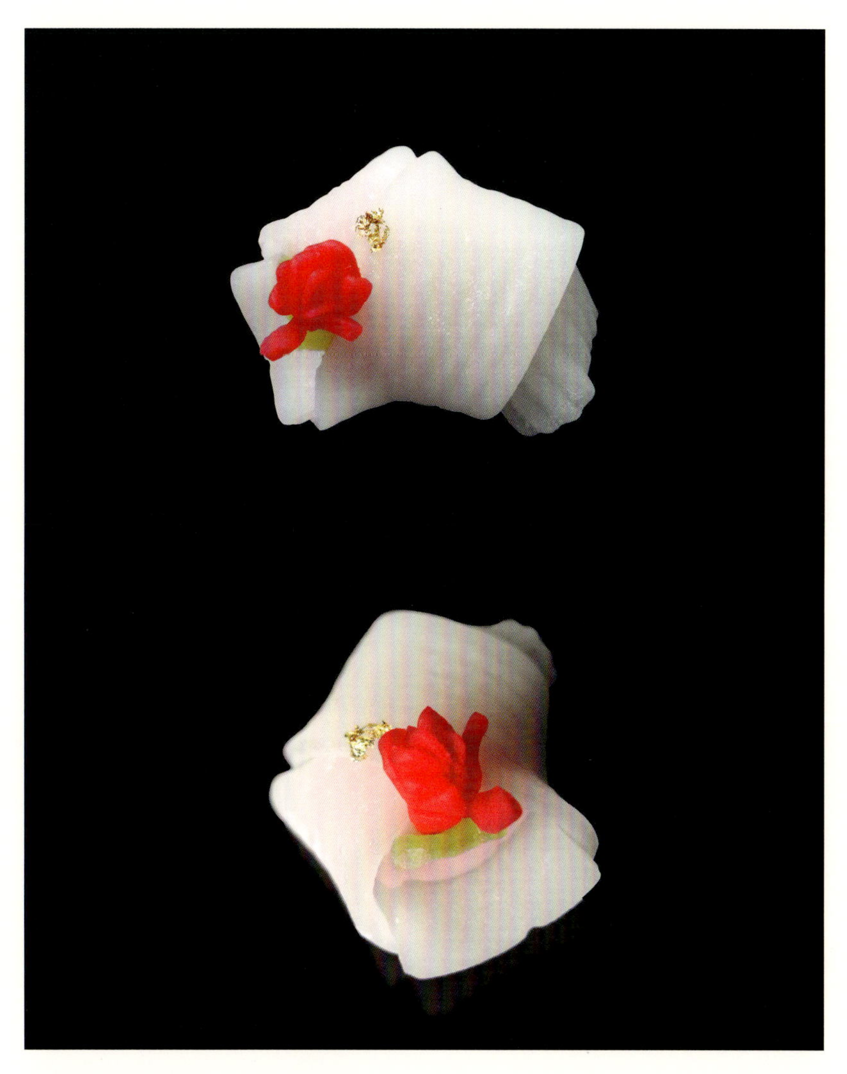

長久堂「ありがとう」 外郎、煉切りあん

Chokyudo “Thanks!” *Uiro, Nerikirian

今日は、母の日。国によって日にちは違うようですね。日は違っても、
おかあさんに感謝する心は同じ。

Today is Mother’s Day. I understand that it is celebrated on different days, depending on the country. The date may differ, but gratitude to mother is the same the world around.

黄色

黄水仙、黄蘗（きはだ）、刈安（かりやす）、黄朽葉（きくちば）、藤黄（とうおう）、芥子（からし）、菜種油、若芽（わかめ）、鬱金（うこん）、支子（くちなし）、檸檬（れもん）、山吹、卵、菜の花、桑……思い浮かぶだけでこれぐらいはある。
仕事では、結構使い分けているかな。
黄色は、有彩色で一番明度が高く明るく、精神の明るさや希望などを表します。明るさから、楽しい、幸せ、願望、自由や解放なども意味します。

長久堂「七重八重」　こなし、赤ごしあん
Chokyudo “Nanae Yae” *Konashi, Akagosian

珍しい黄色いつつじ　Yellow azalea

Yellow: Golden narcissus, Amur corktree yellow, greenish yellow, the color of rotting leaves, gamboge yellow, mustard, rapeseed oil, yellow seaweed, turmeric, gardenia, lemon, *yamabuki*, egg, rape blossom, mulberry: those are all the yellows I can think of.

In my work, each has to be used properly.

The brightest of the chromatic colors, yellow suggests clarity of spirit and the brightness of hope. It also indicates joy, happiness, hope, and freedom.

葵 祭　The Hollyhock Festival

上賀茂神社と下鴨神社の祭礼。
本来は、「宮中の儀」「社頭の儀」と「路頭の儀」の三つで構成されていたが、現在は、「宮中の儀」は省かれ、残り二つだけになってしまっています。葵祭とみなさんが思っているのは「路頭の儀」のこと。本来は、神前で祭文を読み上げ、供物や舞を奉納する「社頭の儀」がメインです。

A festival celebrated by the Kamigamo and Shimogame shrines.

Originally the festival included rituals inside the imperial palace, at the shrines, and in the town. Today, only the latter two are still performed, but what most people think of as the Hollyhock Festival is the "in the town" part, the parade through the city. In fact, however, it is the reading of the address to the deities and the presentation of offerings and dance before them—the rituals performed at the shrines—that are the heart of the festival.

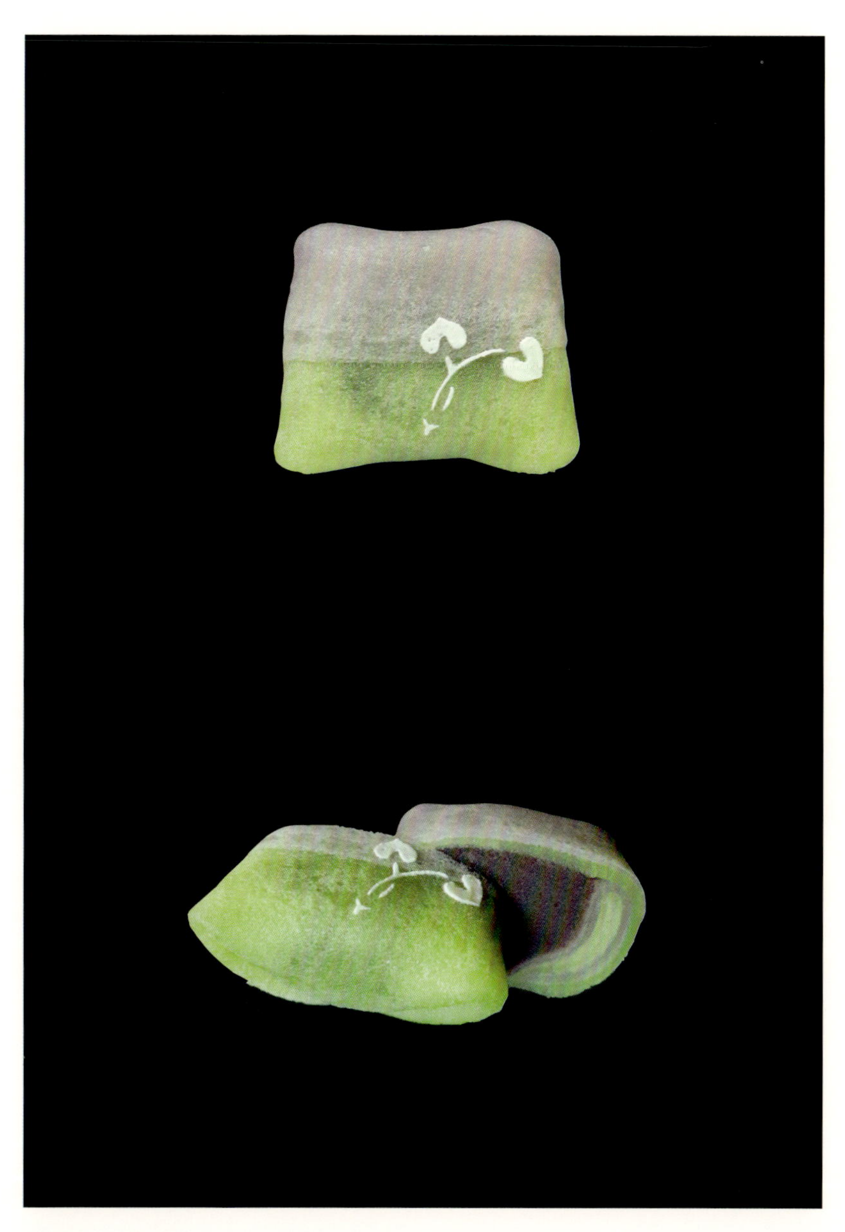

鶴屋吉信「賀茂みどり」 焼皮、黒こしあん
Tsuruya Yoshinobu “Kamomidori” *Yakigawa, Kurokoshian

鶴屋吉信「王朝花傘」　こなし、白こしあん

Tsuruya Yoshinobu “Ocho Hanagasa” *Konashi, Shirokoshian

いまから約1400年前の欽明天皇(539 ～ 571)の時代に大凶作や疫病がまん延した。天皇が占わせたところ、災いは賀茂の神々の祟りであるということが判明。それで天皇が勅使を遣わし、祭礼を行ったのが葵祭(賀茂祭)の起源なんだそうです。
葵祭の名称は祭りの当日、御所内裏の御簾（みす）とか、牛車（ぎっしゃ）、勅使や行列の人らの冠や装束、牛馬などすべてを葵の葉っぱで飾ったことによります。

The origin of the festival goes about some 1,400 years ago, to the reign of the Kinmei Emperor (539 – 571), when the land was plagued with failed harvests and pestilences. The emperor's divinations revealed that these disasters were divine punishment sent by the Kamo deities. He then ordered his envoys to proceed to the Kamo shrines and perform the rites that became the origin of the Hollyhock Festival.

During the festival, the shades on the imperial palace in Kyoto, the ox carts, the headdresses and clothing of the envoys and other people in the processions, even the horses and oxen, are decorated with hollyhock leaves.

葵祭、斎王行列の花笠　The Hollyhock Festival, Hanagasa of Sai-o Gyoretsu

柳の新芽　Willow sprouts

亀屋良長「芽 柳」 煉切り、粒あん
Kameya Yoshinaga “Meyanagi” *Nerikiri, Tsubuan

長久堂「花 筏」　こなし（山芋入り）、赤こしあん

Chokyudo “Hanaikada” *Konashi with yam, Akakoshian

紫野源水「春うらら」　煉切り、白こしあん
Murasakino Gensui “Haru Urara” *Nerikiri, Shirokoshian

紫野源水「本蕨餅」 小豆こしあん
Murasakino Gensui “Honwarabimochi” *Azuki koshian

亀屋良長「花散里」 山芋入りきんとん、こしあん

Kameya Yoshinaga "Hanachirusato" *Kinton with yam, Koshian

長久堂「乙女の唄」　こなし(山芋、小麦粉)、こしあん
Chokyudo “Otome no Uta” *Konashi with yam and flour, Koshian

長久堂「花蓬莱」 雪平、赤こしあん
Chokyudo “Hanahorai” *Yukihira, Akakoshian

長久堂「匂ふ君」 外郎、備中白こしあん
Chokyudo "Nioukimi" *Uiro, Bicchu Shirokoshian

夏
summer

京都鶴屋鶴壽庵「青 梅」 青月餅、こしあん
Kyoto Tsuruya Kakujuan “Aoume” *Aogeppei, Koshian

梅宮大社の梅苑では青梅がなっている。そろそろ梅を漬けるシーズンが近づいてきましたね。
The plum grove at the Umemiya Grand Shrine is full of green plums. Plum pickling season is coming!

塩芳軒「青 楓」 道明寺、粒あん、緑あん
Shioyoshiken “Aokaede” *Domyoji, Tsubuan, Midorian

清水寺の舞台の擬宝珠。この擬宝珠はいろんな説があるのですが、そのひとつにネギ坊主の形というのがあります。

One of the many explanations for these odd bronze fittings atop the railing around Kiyomizu-dera's veranda is that their shape was inspired by an onion.

紫野源水「落とし文」 外郎、白こしあん
Murasakino Gensui “Otoshibumi” *Uiro, Shirokoshian

紫野源水「あやめ」 煉切り、白小豆粒あん

Murasakino Gensui “Ayame” *Nerikiri, Shiroazuki tsubuan

紫野源水「青 楓」 薯蕷、小豆こしあん
Murasakino Gensui “Aokaede” *Joyo, Azuki Koshian

青楓　Aokaede

塩芳軒「ききょう」 外郎、白あん
Shioyoshiken “Kikyo” *Uiro, Shiroan

東福寺に行ったので天得院にも寄りました。いま、桔梗が見頃です。
慶長十九（1614）年、文英清韓長老が住持になったとき、秀頼の請に応じて方広寺の鐘銘を撰文した。しかし銘文中の「国家安康、君臣豊楽」の文字が徳川家を呪詛するものとして徳川家康の怒りを買い、天得院は取り壊され、翌年の大坂夏の陣によって豊臣家は滅ぼされました。後に再建され現在に至っています。

On a visit to Tofukuji temple, I decided to drop by the Tentokuin, and was just in time for the Chinese bellflowers.

Speaking of bells, in 1614, the chief priest of this temple was asked by Toyotomi Hideyori (son and heir of Hideyoshi, the general who had reunified Japan) to select a text for the bell for another temple, the Hokoji. Tokugawa Ieyasu (at the time, Hideyori's guardian) took the text as cursing the Tokugawa clan, to his great displeasure. In the end, the Tokugawa ruled Japan for 250 years, and the Toyotomi were destroyed.

茅の輪くぐり　Passing throush a circle of “Chinowa”

上賀茂神社、ここは和歌を唱えてまわります。
6月22、25日からは、各神社で夏越祓（なごしのはらえ）、年に2回の穢れを祓う行事のうちのひとつ、
神道でいう大祓（おおはらえ）が行われます。
「みな月の　なごしの祓（はらひ）する人は　千年（ちとせ）の命　のぶといふなり」

Here one processes while reciting *waka* verses.

The Nagoshi purification ritual is a major Shinto rite.
There are two such rituals, conducted to remove impurities, each year. “Those who take part in the Nagoshi purification rite in June will extend their lives by a thousand years,” it is said.

亀屋良長「夏 越」 上用、黒こしあん
Kameya Yoshinaga “Nagoshi” *Joyo, Kurokoshian

ここ貴船神社は京都の奥座敷といわれているところ。
京都市内と気温が３～５度は違います。
ここの水無月の大祓では、和歌が３番まであります。
１．みな月の　なごしの祓(はらひ)する人は　千年(ちとせ)の命　のぶといふなり
２．思う事　みなつきねとて　麻の葉を　きりにきりても祓ひつるかな
３．蘇民将来(そみんしょうらい)・蘇民将来

くぐる前に覚えておかないといけません。でもすぐに忘れてしまう。
くぐったら千年も生きられるそうですよ。夏越といえば、京都人は「水無月」ですね。

Here we are at Kibune Shrine.
One of Kyoto's secrets, it is usually three to five degrees cooler here than in the center of the city.
There are three verses to be memorized before passing through the purifying circle of hemp during its Nagoshi purification ceremony:

1. Those who take part in the Nagoshi purification rite in June will extend their lives by a thousand years.
2. Cut those hemp leaves and cut them again in June, but will you be purified?
3. Somin Shorai, Somin Shorai (lucky charm, lucky charm)
But I always forget them immediately.
They do say going through this purification in June (*minatsuki*) brings long life.

貴船神社夏越大祓ちのわのお札　Kibune Shline Nagoshi Oharae Chinowa

京都鶴屋鶴壽庵「水無月」 外郎
Kyoto Tsuruya Kakujuan “Minazuki” *Uiro

上等の水無月は値段が普通の倍はしますが、やっぱり値段だけのことはありました。
右）これが一番ポピュラーな形です。京都人は、これを食べないといけないと思っています。
This confectionary is a deluxe form of Minatsuki. It is twice as expensive as usual, but well worth it. It’s also the most popular version. In Kyoto, you can’t do without it.

塩芳軒「水無月」 外郎 / 長久堂「水無月」 外郎、黒糖
Shioyoshiken “Minazuki” *Uiro / Chokyudo “Minazuki” *Uiro Kokuto

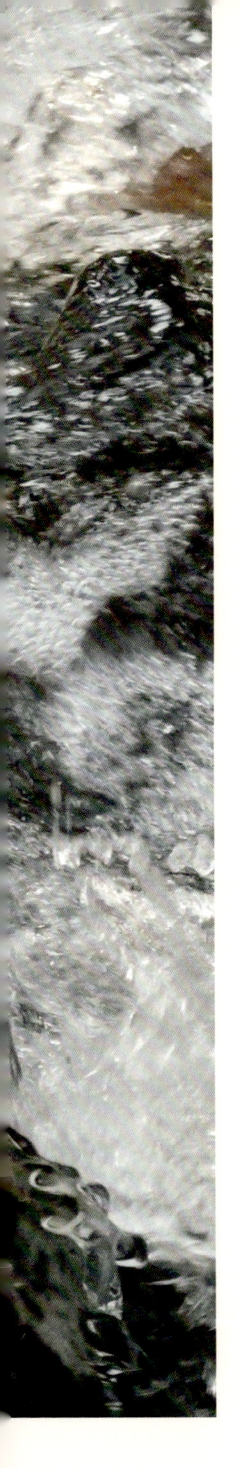

水無月（みなづき）は、「水の無い月」と書きますが、水が無いわけではない。「無」は、神無月と同じように、「の」にあたる連体助詞「な」で、「水の月」という意味です。旧暦6月は田んぼに水を引く月であるから、水無月といわれるようになりました。京都では、「水無月」というお菓子を食べます。
もともとは、冬に氷室で氷を保存し、旧暦の6月1日に氷を取り出し、宮中に献上した。献上された氷は、帝が暑気払いに小豆をのせて「あまかずら」という植物の蔓を煎じた甘汁をかけて食べたんだそうです。
京都の氷室は、西賀茂と丹波にありました。メインの氷室は西賀茂で、丹波は予備だったそうです。 いまでも西賀茂の氷室は、京見峠（きょうみ）の北東、現在の氷室町にあります。

氷などは、庶民の口には入らないから、白の生地に小豆をのせ、三角形に包丁された菓子にしました。水無月の上部にある小豆は悪霊祓いの意味があり、三角の形は暑気を払う氷を表しているといわれています。それさえ庶民が食べられるようになったのは、明治以降のこと、この水無月、他の地域に行くと食べない、というか、ないことに京都人はびっくりする。
「水無月」という名前は、京都府菓子工業組合の登録商標なんだそうです。だから地方で勝手に作って売れないというのが真相のようです。

The traditional name for the sixth lunar month, *minatsuki*, means “water month” and refers to the custom of flooding the rice paddies at this time of year. In Kyoto, this is the season for *minatsuki* sweets.

Long ago, ice would be stored in an icehouse in the winter, then taken out on the first of the sixth lunar month and offered to the imperial court. Azuki beans and a sweet liquid made by boiling the *amakazura* vine were poured over the ice, which the emperor then ate to forget the summer’s heat. The Kyoto icehouses were in the West Kamo and Tanba areas; West Kamo was the site of the main icehouse and Tanba of its backup. The West Kamo icehouse lives on in the geography of the city today in the place name Himurocho (“icehouse district”) to the northeast of Kyomitoge Pass, in the northern part of the city.

Ordinary people did not, of course, eat anything as exotic as ice. Instead, we invented *Minatsuki*, a confection in which a white sheet of *uiro* (a gelatinous ingredient made of glutinous rice flour or *kuzu*) is covered with beans and cut into triangles. The beans imply driving out evils spirits, while the triangular shape is said to express the idea of ice, thus driving out the heat. While *Minatsuki* did not become popular until the Meiji period, people from Kyoto are often surprised to learn that it is not available in other parts of the country. (The truth of the matter seems to be that *Minatsuki* is a registered trademark of the Kyoto Confectionary Industrial Guild.)

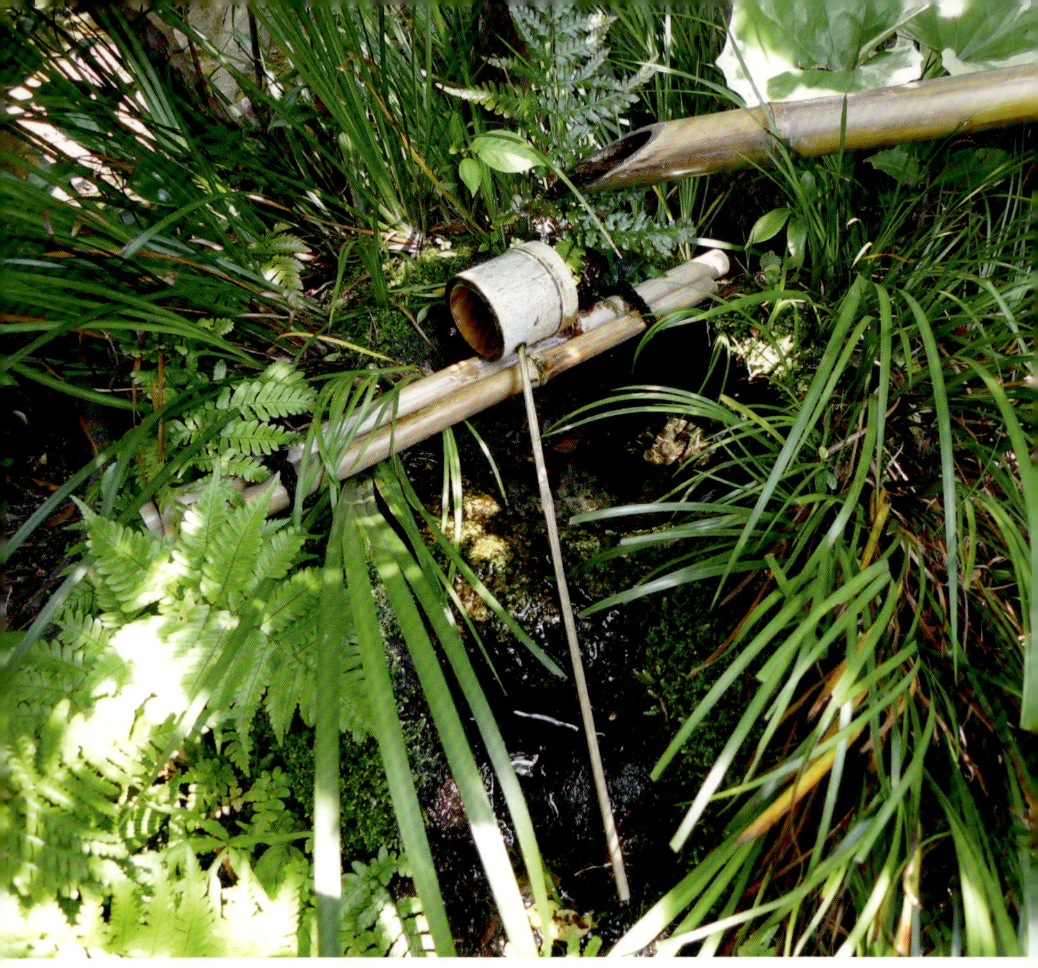

水琴窟(すいきんくつ)の音、涼しかるらん

Suikinkutsu The subtle sound of water dripping into a buried jar Clarify Coolness

水琴窟の音が妙心寺退蔵院で聴けます。
本物を聴く機会はそうそうありません。
枯山水の庭を抜けると本物の池があります。
ありました。
よく耳を澄ますと聴こえました。

To hear the real thing, visit the Taizoin, a temple in the Myoshinji complex.
My opportunities to hear it are few. Yesterday, however,
I was at the Taizoni, went past the dry landscape garden, and reached an actual pond.
There it was.
Listening intently, I heard it.

長久堂「水琴窟」 黒糖吉野葛、赤ごしあん玉、備中鹿の子豆入り

Chokyudo "Suikinkutsu" *Kokuto Yoshino kuzu, Akagoshian, Bicchu kanokomame

鍵善良房「氷 室」 煉切り、粒あん
Kagizenyoshifusa “Himuro” *Nerikiri, tsubuan

亀屋清永「渓 流」 吉野葛、白こしあん
Kameya Kiyonaga "Keiryu" *Yoshinokuzu, Shirokoshian

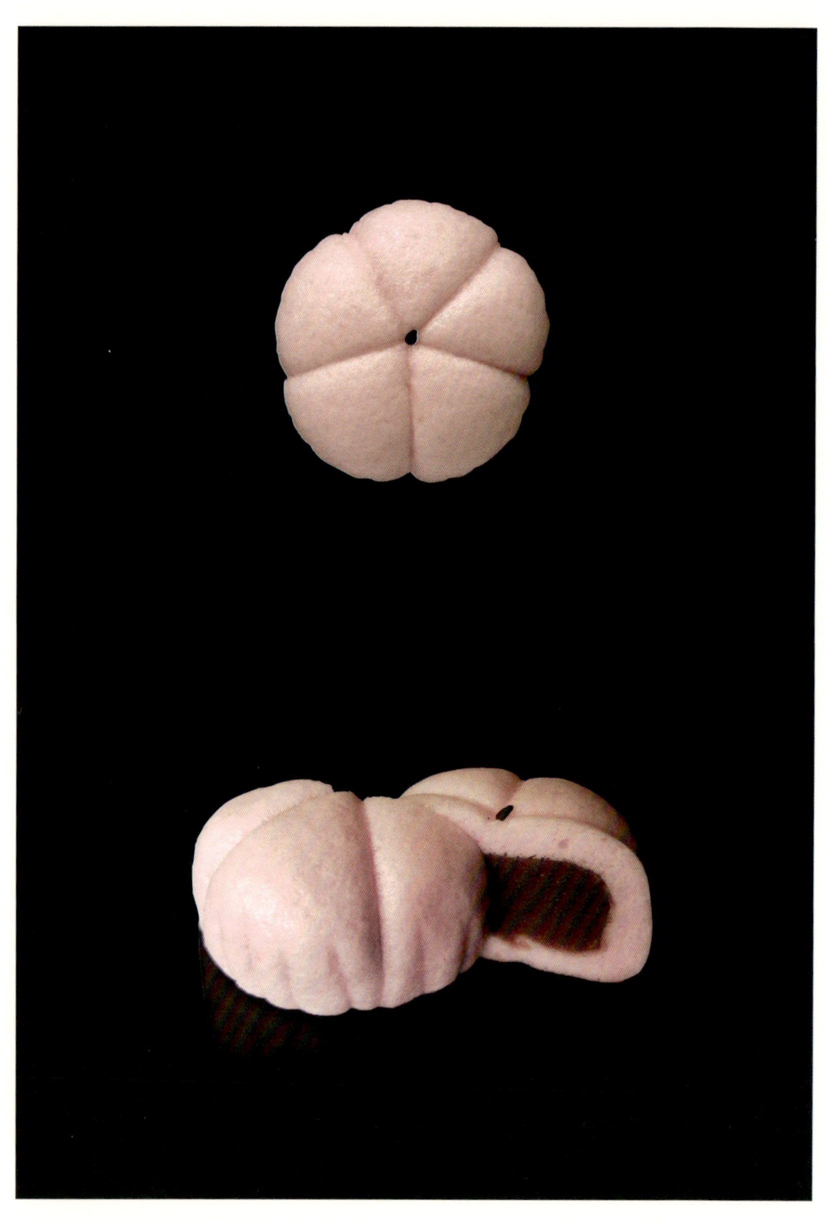

二條若狭屋「なでしこ」 上用、こしあん
Nijo Wakasaya “Nadeshiko” *Joyo, Koshian

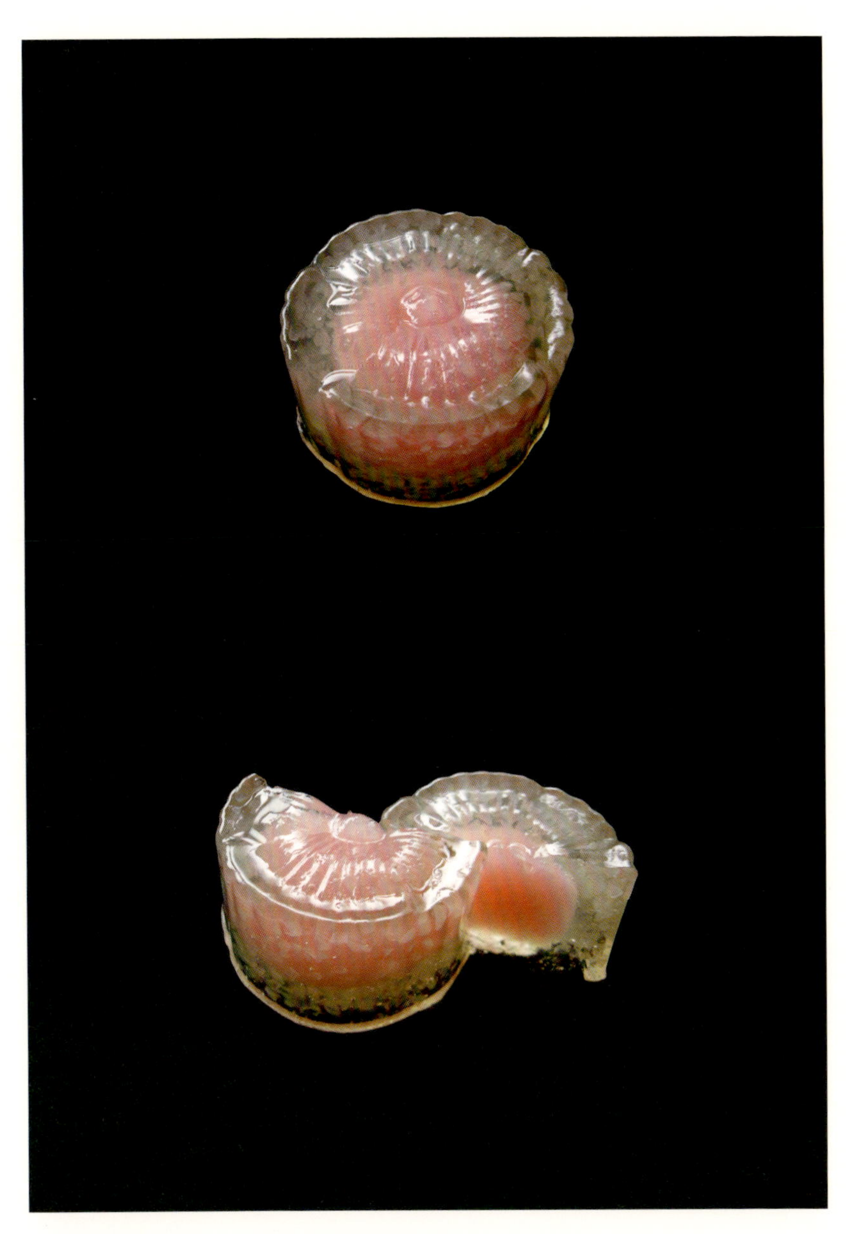

千本玉壽軒「撫子」 道明寺入り羊羹、赤色白こしあん
Senbon Tamajuken “Nadeshiko” *Yokan with Domyoji, Akairo Shirokoshian

撫子(なでしこ)って、見かけそうでなかなか見かけない。
名前の由来は、我が子を撫でるようにかわいがるのと同じように、
かわいい花ということで。「撫子」という名前になったんだそうです。

Flowering pinks seem strangely elusive.
Their name in Japanese, *nadeshiko*, refers to “stroking a child,” because the flowers are so charming one yearns to stroke them.

本願寺唐門　Honganji Karamon

大徳寺土壁　Daitokuji Tsuchikabe

2時ぐらいに起きると、突然鳥が鳴いた。
「ほととぎす　鳴きつる方を　ながむれば　ただありあけの　月ぞ残れる」
後徳大寺左大臣（ごとくだいじのさだいじん）

ホトトギスは、夏の到来を知らせる鳥。平安時代から
「夏を告げる鳥」「夜中に鳴く鳥」だったそうです。もちろん昼間も鳴くだろうけど。
平安時代の人は、夏のホトトギスの最初の声「初音」を聞くために
一晩中起きていたというようなことをしたんだそうです。
ひょっとしてこの間のは、「初音」だったのかな……。
ならちょっと得をしたかも。

When I woke up at about 2 a.m., suddenly a bird sang.

"Looking in the direction of the cuckoo cry, I see only the moon as dawn approaches." — Gotokudai no Sadaijin

The Japanese cuckoo, or *hototogisu*, heralds the arrival of summer. It has been known since Heian times as the harbinger of summer or the bird that sings at night—though it does of course sing in daytime, too.

In Heian times, people would stay up all night to hear the first cry of the cuckoo each year, they say. Perhaps what I heard was that first cry—lucky me!

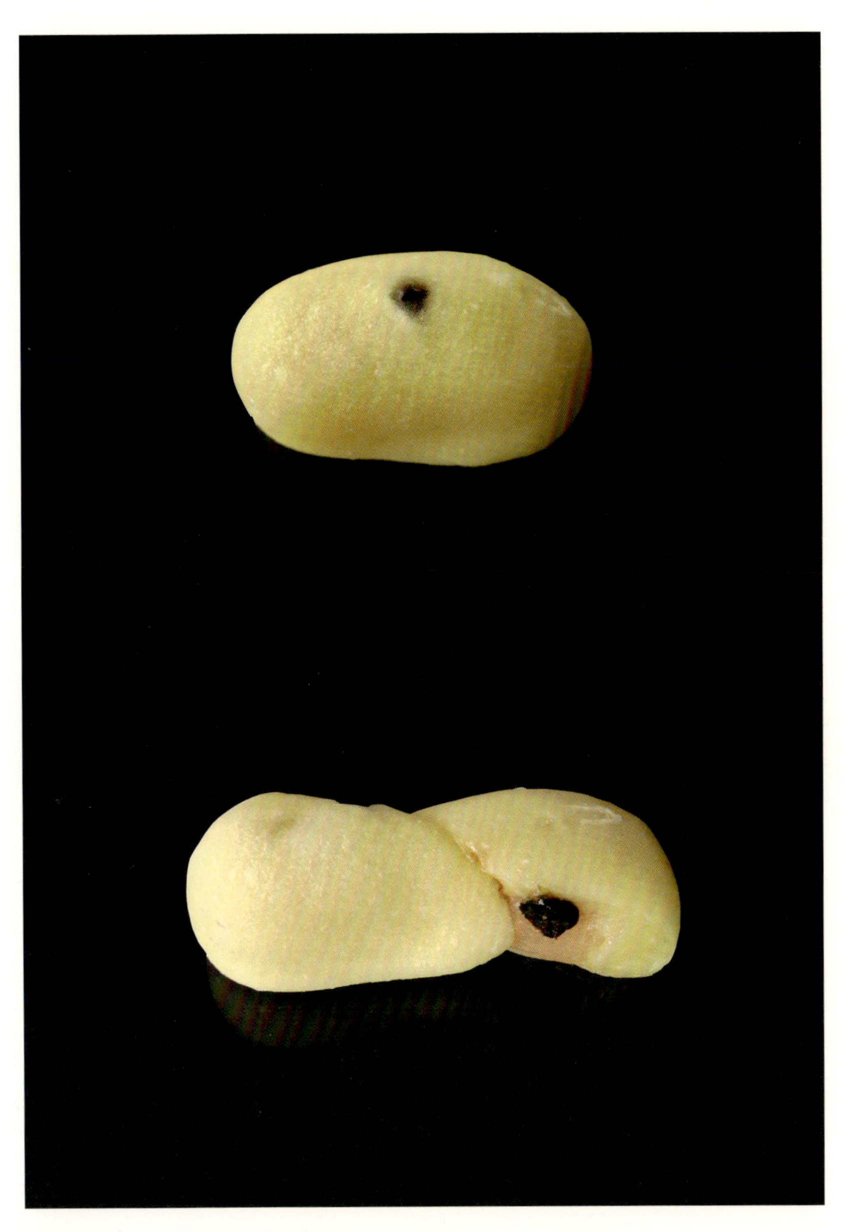

紫野源水「一声（ほととぎす）」 羽二重餅、白こしあん、大徳寺豆
Murasakino Gensui “Chirp (hototogisu)” *Habutaemochi, Shirokoshian, Daitokujimame

そろそろ紫陽花が見頃になってきた。
梅雨入りももうそろそろでしょうか。
うっとうしいけれど雨も降ってもらわないとまた水不足になる。
京都の場合は、水不足にはならないけど、
琵琶湖に藻が異常発生して水道水がカビ臭くなるのがかなわない。
紫陽花にはさまざまな色がありますが、一本の紫陽花が咲いてから枯れるまでもさまざまにその色を変え、一名に「七変化」とも呼ばれます。そこから、紫陽花は変わりやすい心の比喩に用いられました。心変わりや移り気、あるいは冷淡な性格を象徴させることがある花です。不変の価値を尊ぶのはおそらく民族の気質として古代から変わらないものなのでしょう、紫陽花があまり良くないイメージを負いはじめたのは、相当古い時期からのことと考えられます。

It is almost time for the hydrangeas to bloom—and for the rainy season to begin.

The season is rather gloomy, but without rain, we will have another water shortage. Kyoto itself may not be short of water, but Lake Biwa might suffer an algae bloom, and the tap water would be horribly smelly.

Hydrangeas, turning to a more cheerful topic, come in various colors, and the color on one plant may change as the flowers bloom and fade.

In fact, the hydrangea is often used as a metaphor for someone whose affections change easily. Flighty, inconstant, indifferent: those are the characteristics the hydrangea symbolizes. Given the Japanese predilection for revering unchanging values, it seems likely that the hydrangea has born the burden of that rather negative image for a long, long time.

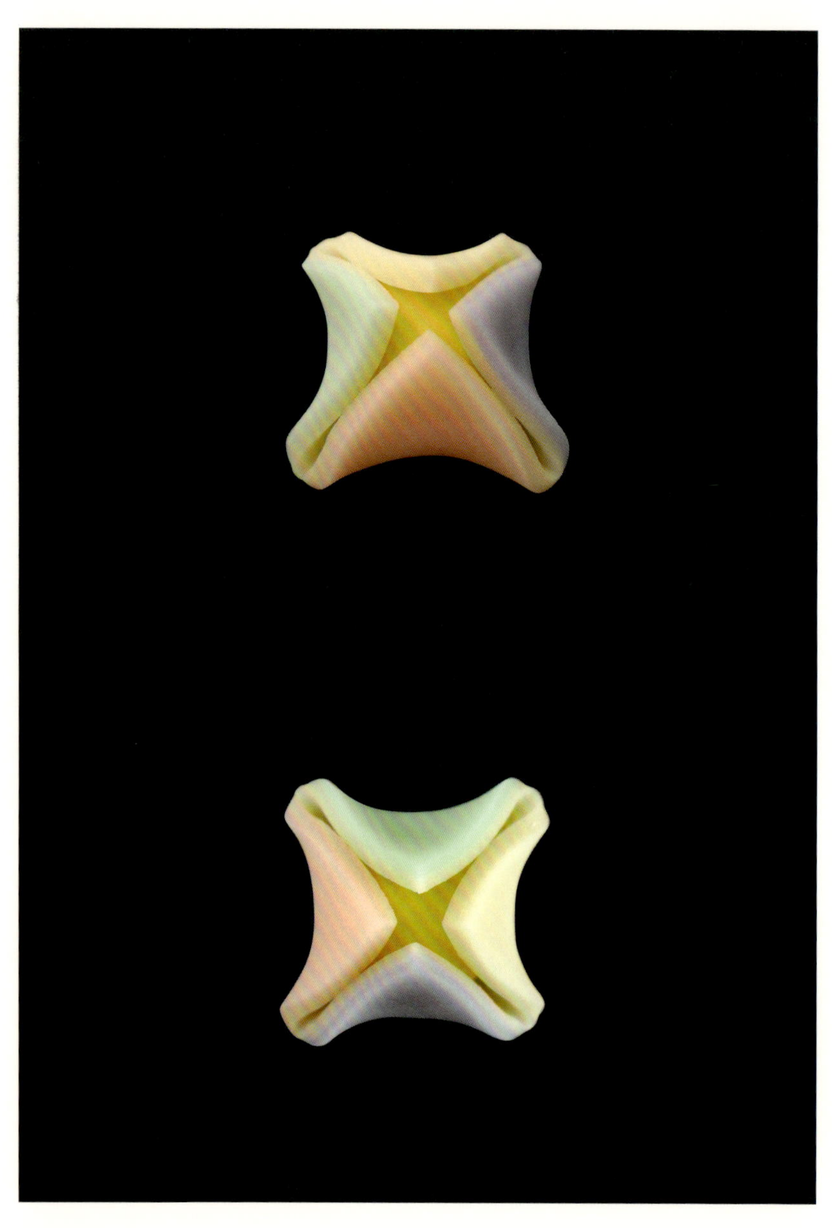

長久堂「よひら」　こなし、白こしあん
Chokyudo “Yohira” *Konashi, Shirokoshian

私は、ガクアジサイのほうが好き。
この四弁の花と見られてきたものは実は花ではなく萼で、
紫陽花の本当の花は中央に集まる小さな粒のような部分なんです。

I prefer lacecap hydrangeas.
What look like four-petalled flowers are actually showy bracts.
The actual flowers are the little buds in the center.

亀屋良長「四ひら」 煉切り、白こしあん
Kameya Yoshinaga "Yohira" *Nerikiri, Shirokoshian

京都鶴屋鶴壽庵「紫陽花」 葛、黄身あん
Kyoto Tsuruya Kakujuan “Ajisai” *Kuzu, Kimian

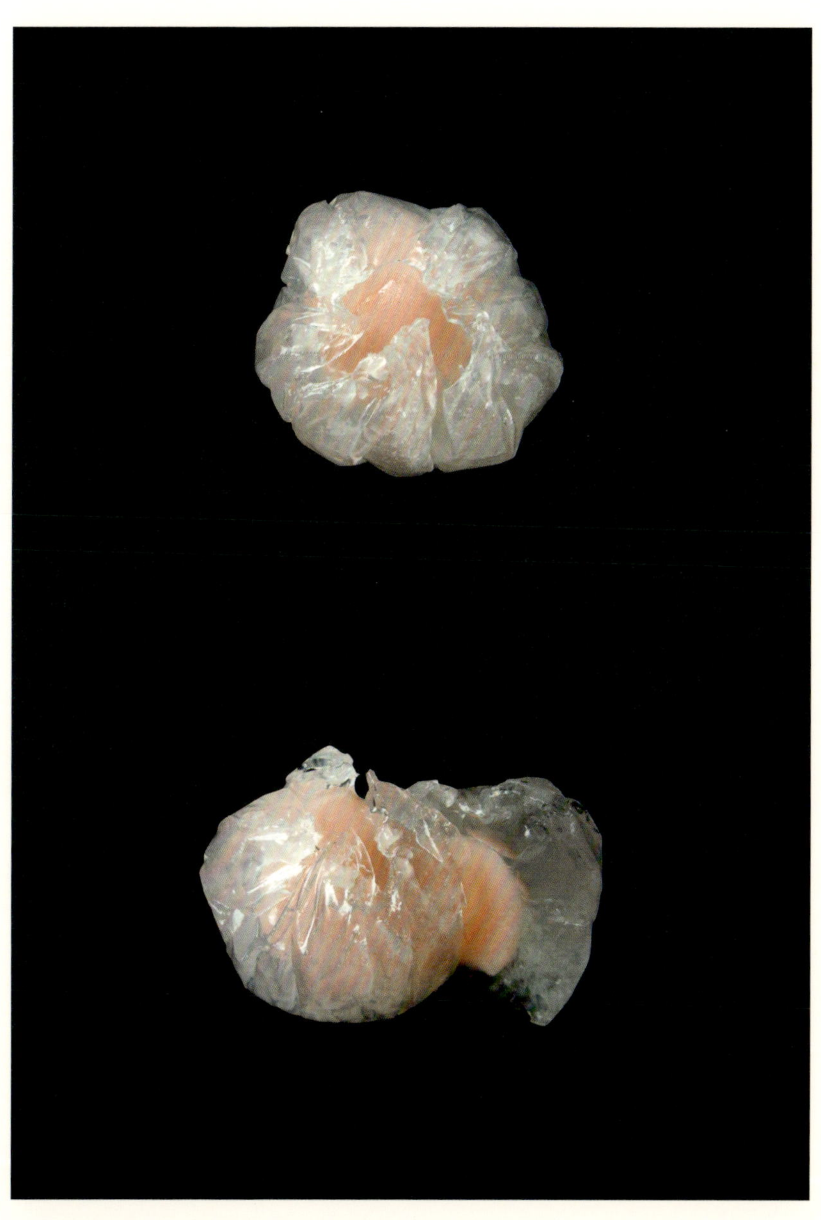

総本家駿河屋「水ぼたん」 錦玉、道明寺、白こしあん
Souhonke Surugaya "Mizubotan" *Kingyoku, Domyoji, Shirokoshian

やっと梅雨らしい空模様になってきた。
紫陽花は球状のセイヨウアジサイと、日本原産のガクアジサイがあります。
花の色は、アントシアニンや発色に影響する補助色素や土壌の酸性度や
アルミニウムイオンの量で変わるそうです。
単純に土が酸性なら青でアルカリ性なら赤というものではないそうです。
初めは青く咲いていて、咲き終わりに近づくにつれて赤みがかってきますよね。

Yes, it looks like the rainy season is here.
Hydrangeas come in the mophead type, introduced from Europe, and the lacecap type, native to Japan. Their color varies, due to an anthocyanin and supplementary color elements that effect color production, depending on the acidity of the soil and its aluminum content.
It's apparently not as simple as acid soil producing blue flowers and alkaline soil turning them pink. Some hydrangeas are blue when the flowers first appear, then gradually turn reddish as their season nears an end.

長久堂「七変化」 みじん羹、赤こしあん
Chokyudo "Shichi henge" *Mijinkan, Akakoshian

長久堂「瑞居の風」 こはくかん、卵白入り

Chokyudo "Zuiko no Kaze" *Kohakukan, egg white

涼をもとめるといえば、音なんかも大事ですね。
蹲（つくばい）に水が落ちる音や鹿威し（ししおど）、水琴窟（すいきんくつ）なんかの音もいい。
日本の庭は、うまくつくられていると思う。
瑞は「めでたい」という意味。
居は、すわること。いること。また、その所。（接尾語的に）存在すること。存在する所。
そこに居てるだけで、めでたいってなんかわかるような気がする。
なんか見ているだけで、涼しく感じられます。
琥珀羹（こはくかん）は、冷やさないで食べるとちょっと甘すぎ。
冷やすことが前提で甘味を調整してあるんだろうな。
葛と違い、冷蔵庫に入れっぱなしにしても濁らないから管理も楽だし。

In the quest for a refreshing sense of coolness, sounds can make a big difference. The sound of water falling into a basin, of a water-filled bamboo tube clacking against a stone, even of water dripping into a buried jar can help. Yes, Japanese gardens are well designed.

The title suggests what a blessing a breeze can be. This confection communicates coolness at a glance.

Since amber agar-agar is overly sweet unless chilled, this sweet should be prepared on the assumption that it will be served chilled. Fortunately, unlike *amazura*, amber agar-agar-based sweets will not spoil if left in the refrigerator. That too is a blessing.

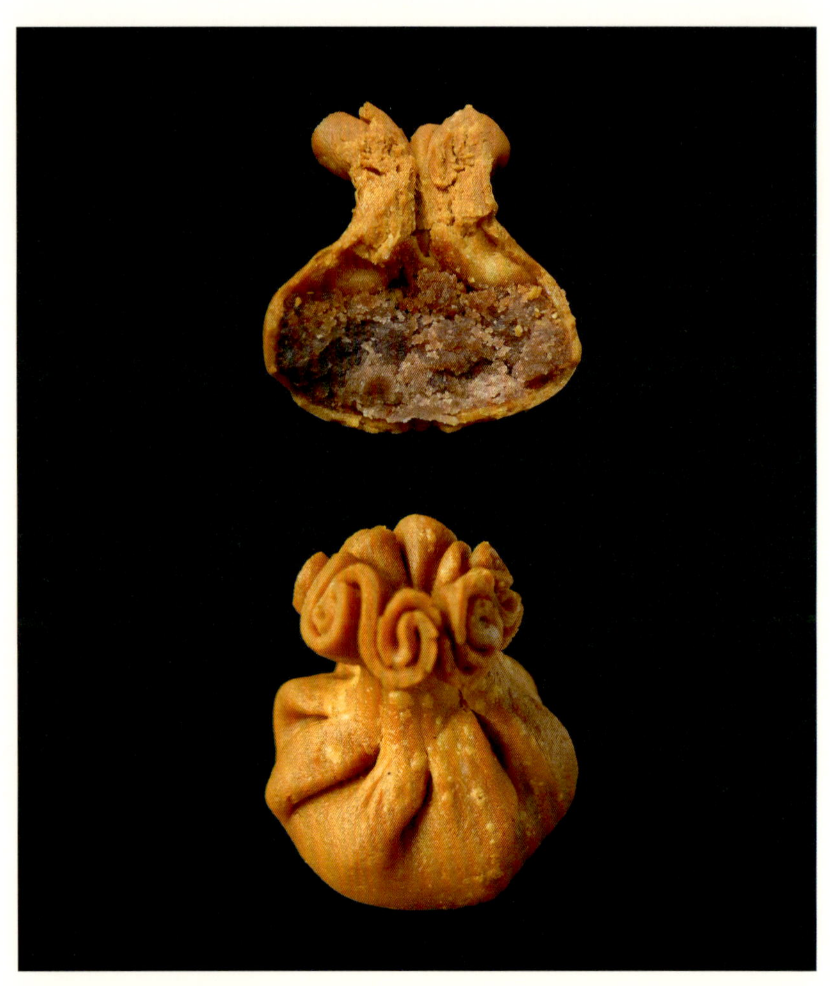

亀屋清永「清浄歓喜団」
Kameya Kiyonaga “Seijokankidan”

略してお団と言い、遠く奈良時代遣唐使により我国に伝えられた唐菓子の一種で、数多い京菓子の中で、千年の歴史を昔の姿そのまま、今なお保存されているものの一つであります。

Known as just “O-dan,” these are an example of *karagashi*, confections that were introduced to Japan by the envoys sent to China long ago, in the Nara period (eighth century). Among all the varieties of *wagashi* made in Kyoto, O-dan are the only one that has a history of over a millennium and is still made, just as it was back then.

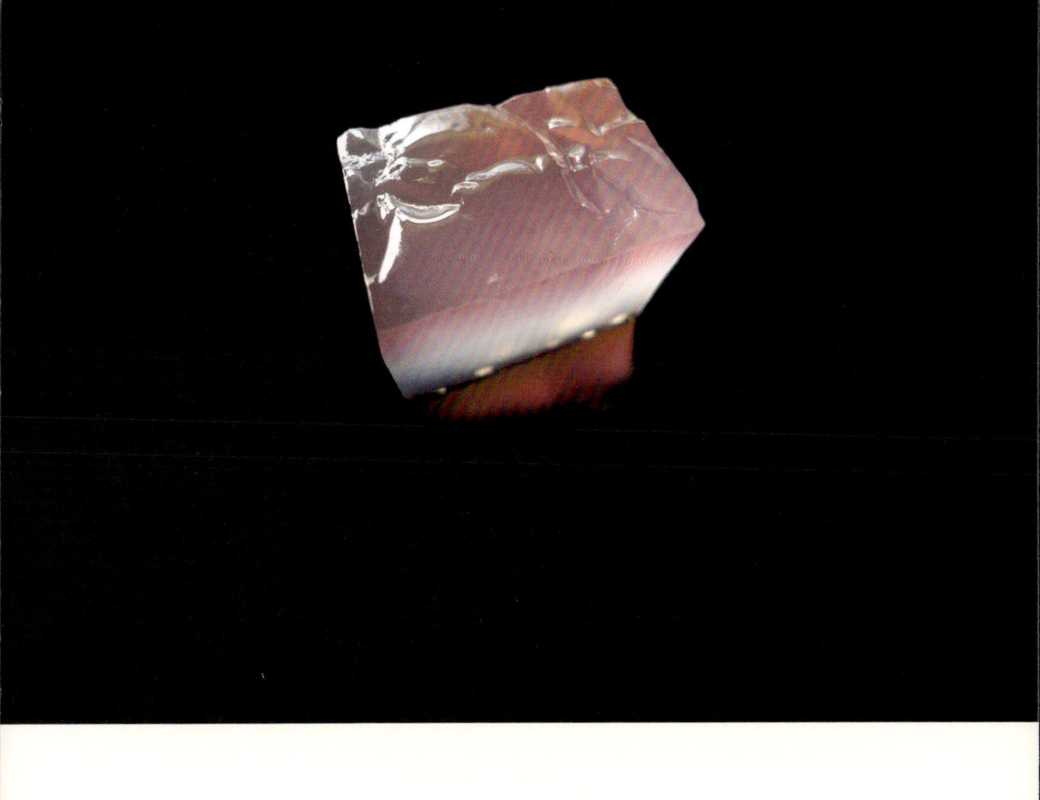

総本家駿河屋「夕立」　錦玉、葛
Souhonke Surugaya “Yudachi” *Kingyoku, Kuzu

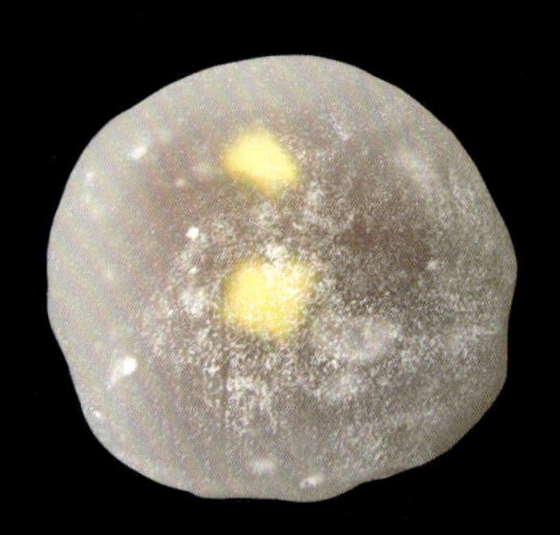

千本玉壽軒「星月夜」 葛、黒こしあん
Senbon Tamajuken “Hoshizukiyo” *Kuzu, Kurokoshian

この上生菓子は葛製なのでとても肌触りがいい。
生菓子の中に宇宙を感じる不思議な感覚です。

The use of *kuzu* in this deluxe moist sweet gives it a delicate texture.
What a strange sensation, to perceive the universe within a sweet

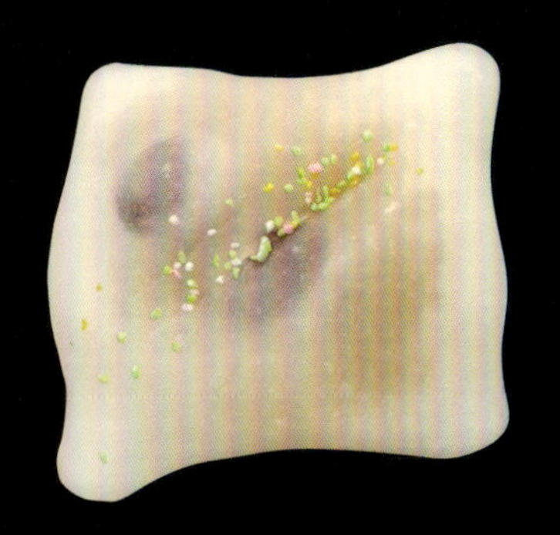

紫野源水「天の川」 外郎、白小豆あん
Murasakino Gensui “Amanogawa” *Uiro, Shiroazukian

京都鶴屋鶴壽庵「星の光」　葛、黄あん
Kyoto Tsuruya Kakujuan “Hoshino Hikari (Starlight)” *kuzu, kian

亀屋良長「星に願いを」　琥珀、白あん

Kameya Yoshinaga “Hoshi ni Negai wo (Wish upon a star)” *Kohaku, Shiroan

今日は、七夕。
短冊などを笹に飾る風習は、夏越(なごし)の大祓(おおはらえ)に設置される茅の輪の両脇の笹竹に因んで江戸時代から始まったもので、日本以外では見られません。七夕は、本来7月7日に行われていた神迎えの儀式に、中国の行事が結びつき、彦星と織姫の物語とあいまって民間に広がりました。神迎えの儀式とは、水辺に棚（祭壇）を設え、その棚に神の衣を織って奉上する儀式。この衣を織る機織を棚機(たなばた)というのです。衣を織る乙女を乙棚機(おとたなばた)といった。その乙女が神さんを迎える。穢(けが)れを祓ってもらいたい人たちが川で禊(みそぎ)を行い、神さんにその人たちの穢れを持ち去ってもらうというもの。この儀式が七夕の元なんです。

※ 禊：神道で自分自身の身に穢れのあるときや重大な神事などに従う前に、自分自身の身を川や海で洗い清めること。

Today is Tanabata, the Star Festival.

The Tanabata custom of adorning bamboo branches with poem slips and other decorations is found only in Japan. It goes back to the Edo period, and is based on the association with the bamboo branches at both sides of the ring through which people passed during the summer purification rites. Tanabata itself, however, originated as Chinese festival for welcoming back the gods on the seventh of the seventh lunar month; in the Edo period, it became widespread among the common people because of its association with the tale of the Weaver Maid and the Cowherd.

The annual event that is the origin of Tanabata lacked the decorations we associate with it today. In a ritual to welcome the gods, an altar (*tana*) was set up by a stream and a garment for the gods woven atop it. The loom used to weave that garment was called the *tanabata*, the origin of the name of the festival; the maiden weaving with it, the *ototanabata*, was welcoming the gods. People wishing to be cleansed of their impurities would carry out a purification rite in the river, and the deities would carry away their impurities.

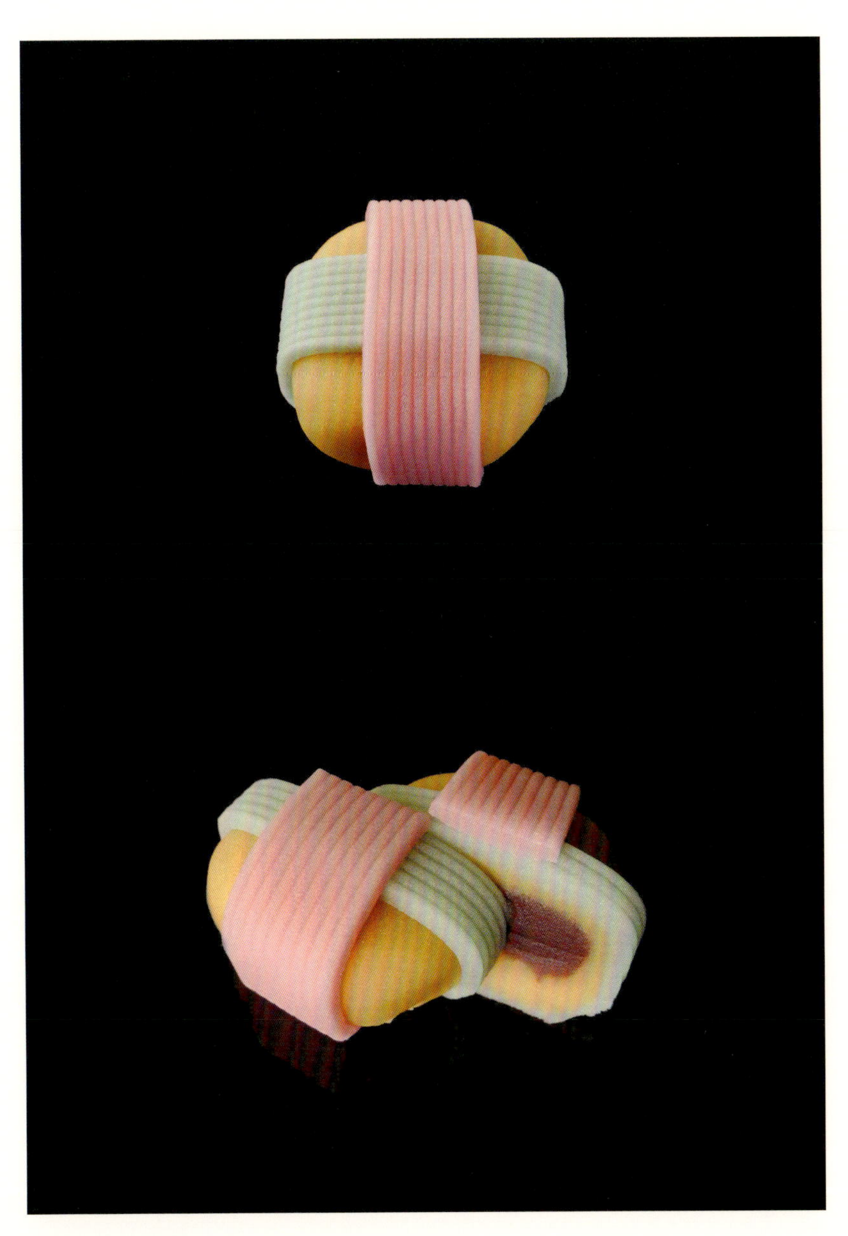

亀屋良長「織 姫」 煉切り、黒こしあん
Kameya Yoshinaga “Orihime” *Nerikiri, Kurokoshian

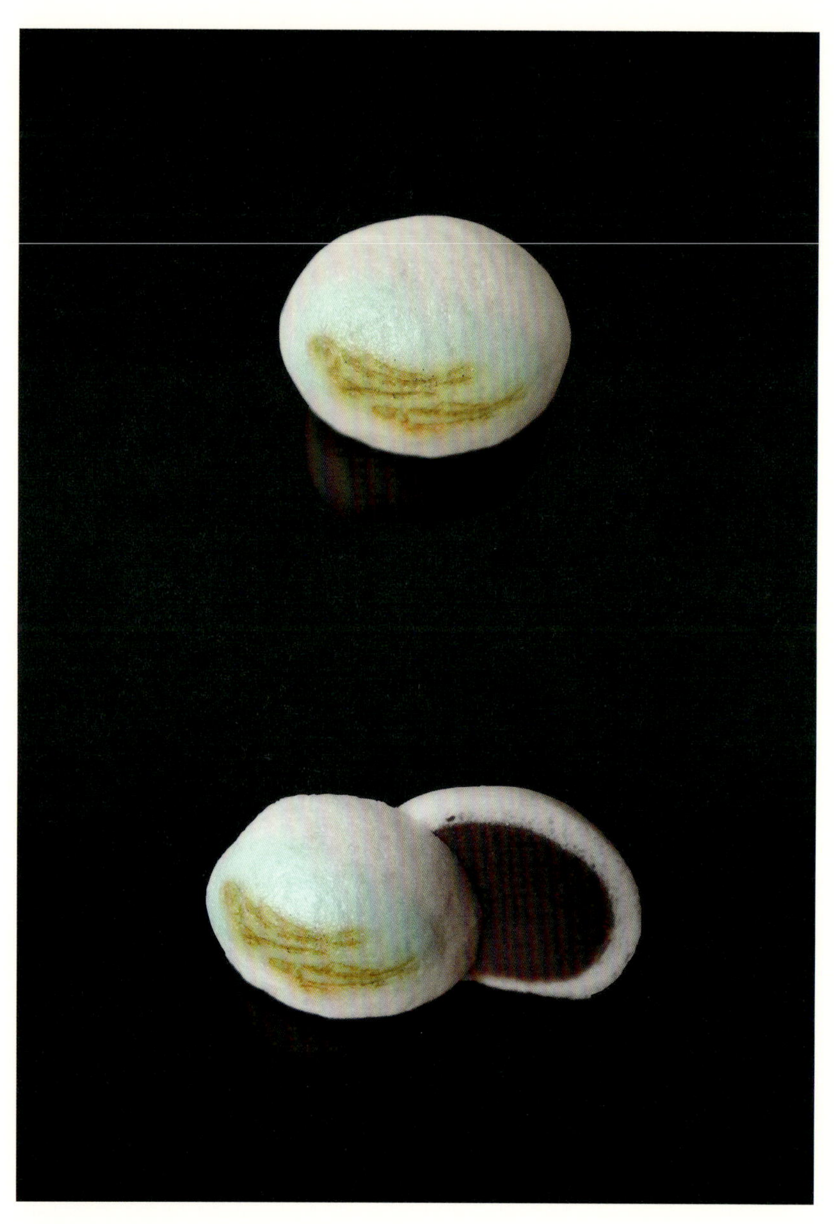

鍵善良房「若 鮎」 上用、こしあん
Kagizen Yoshifusa “Wakaayu” *Joyo, Koshian

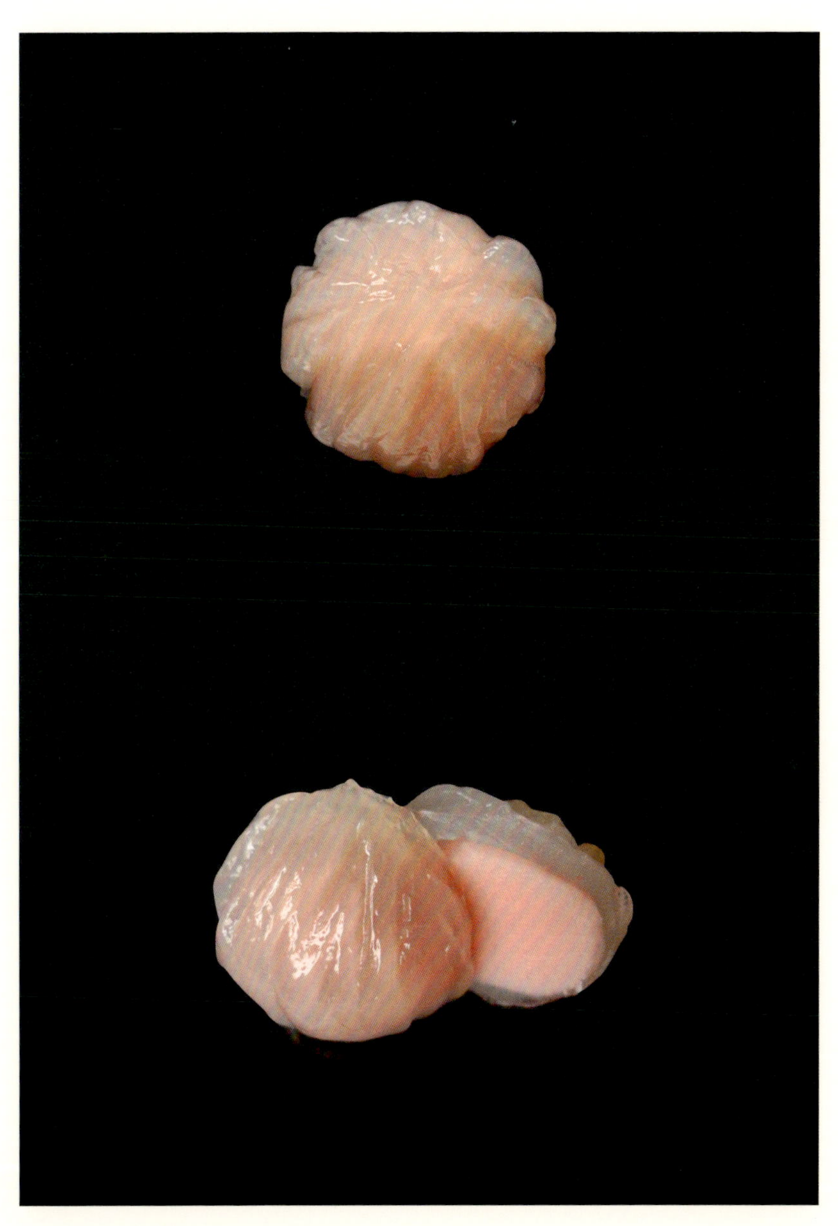

塩芳軒「水ぼたん」　葛、しろあん
Shioyoshiken “Mizubotan” *Kuzu, Shiroan

紫野源水「藻の花」 琥珀羹
Murasakino Gensui “Mo no Hana” *Kohakukan

じゅんさい入り、こんな和菓子は初めて見ました。
Look at topped Junsai, I've never seen such *Wagashi* before.

亀屋良長「水の音」 琥珀、道明寺
Kameya Yoshinaga “Mizu no Oto (Splash)” *Kohaku, Domyoji

紫野源水「朝露」　きんとん、小豆粒あん
Murasakino Gensui “Asatsuyu” *Kinton, Azuki tsubuan

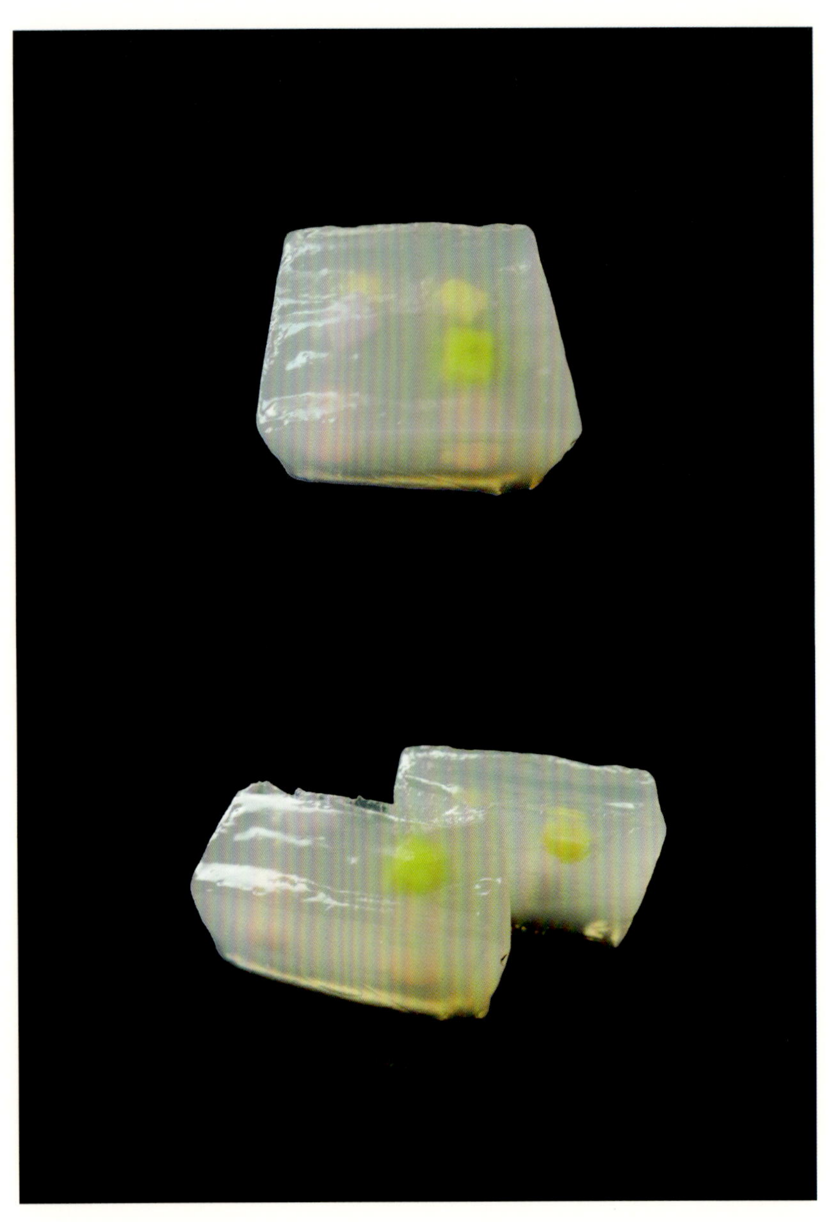

長久堂「花 氷」 葛
Chokyudo “Hanagori” *Kuzu

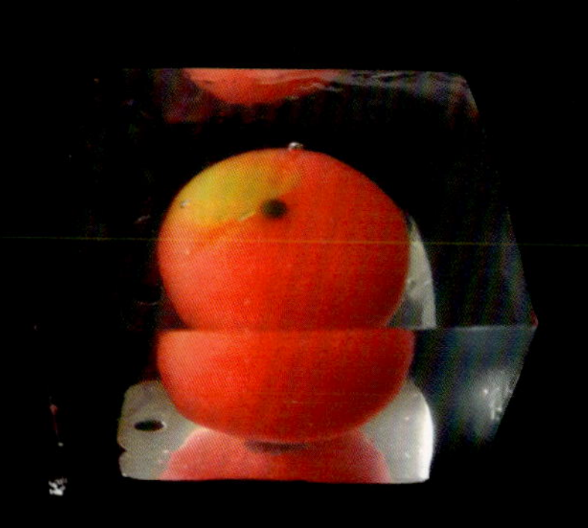

長久堂「夏惜しむ」 琥珀、備中白こしあん
Chokyudo "Natsu Oshimu" *Kohaku, Bicchu shirokoshian

上生菓子は、そろそろ華やかになってきました。実際の季節より確実に秋をたのしめそう。
Deluxe moist sweets are now becoming more colorful.
It looks like we are in for a foretaste of fall.

二條若狭屋「夏の山」 上用、黒こしあん
Nijo Wakasaya “Natsu no Yama” *Joyo, Kurokoshian

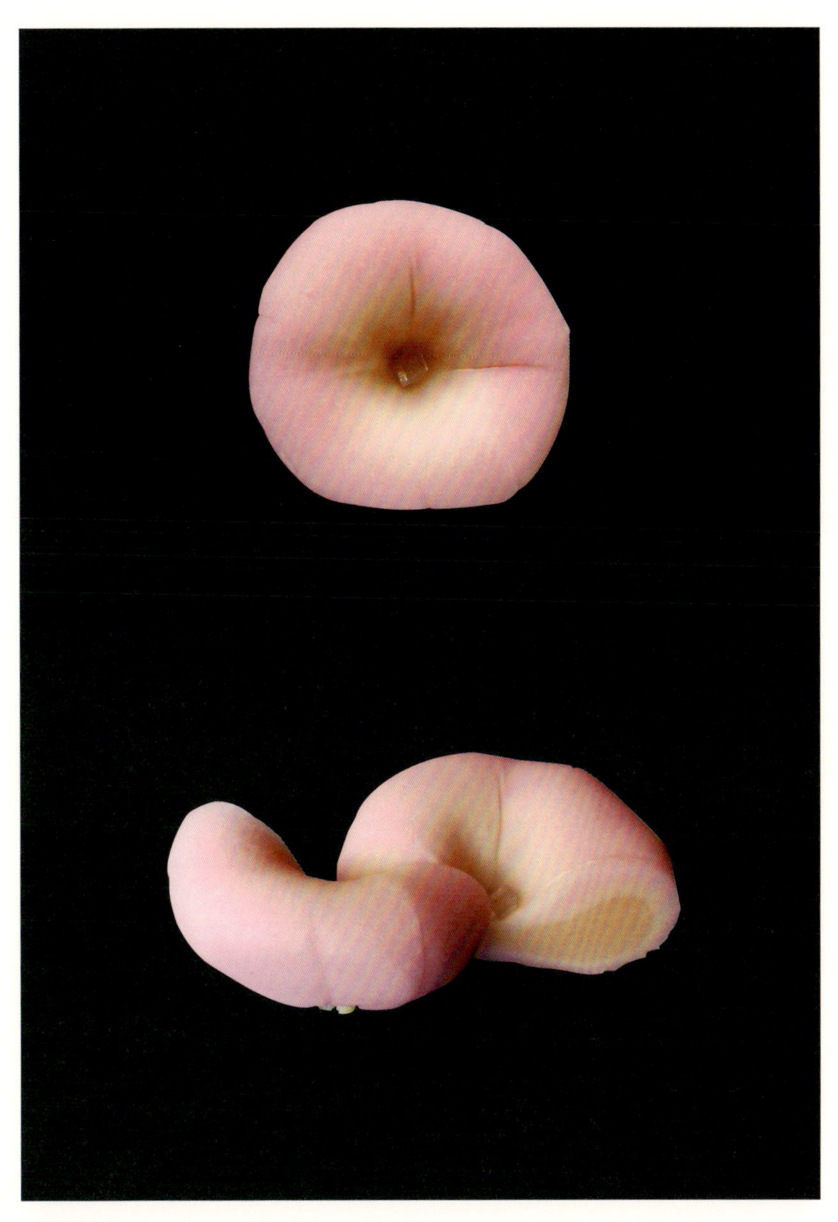

紫野源水「もらい水」　煉切り、白小豆こしあん
Murasakino Gensui “Moraimizu” *Nerikiri, Shiroazuki koshian

やっと京都も梅雨明けになったようです。
明けたと思たら週末は、もう六道詣り。 お精霊（しょらい）さんを迎えに行かないけません。
もうお盆ですよ、言うてる間に送り火になってしまう。

The rainy season is over at last, even in Kyoto.
Summer weather: that means this weekend is already time for *Rokudo mairi*, the temple visits made in preparation for Obon, for welcoming the spirits of our ancestors.
Yes, the Obon season is here.
Before I know it, the *okuribi*, the symbolic fires on the mountainsides to guide the spirits back again, will be lit.

末富「京五山」 懐中ぜんざい
Suetomi “Kyogozan” *Kaichuzenzai

京都五山送り火　Kyoto Gozan no Okutibi

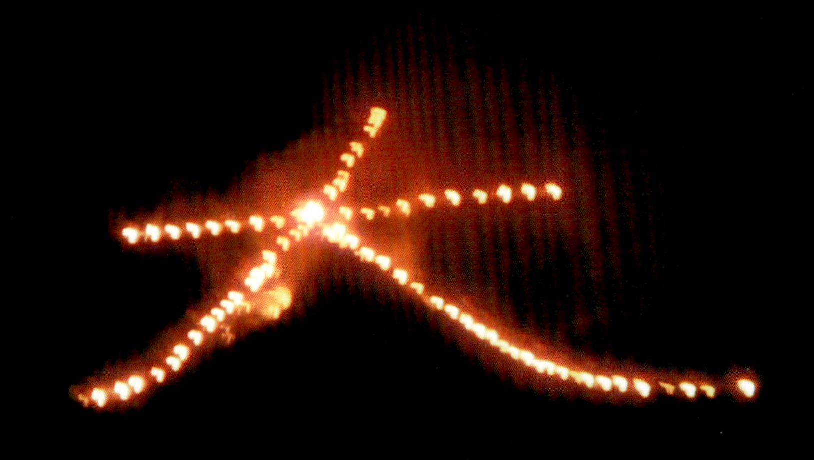

8月16日 大文字の送り火 Daimonji no Okuribi 16th of August

ついこの間、六道詣りだと思っていたら、今日はもう「五山送り火」。
京都のお餅屋さん、おまん屋さんの店頭では、お盆用にお供えのお菓子を売ったハリます。
お供えは日によってかわります。
１２日／白餅、お迎えだんご、はす菓子
１３日／白餅、お迎えだんご、おはぎ、はす菓子
１４日／白餅、白むし（白いおこわ）、おはぎ
１５日／白餅、送りだんご、白むし
１６日／送りだんご、白餅
これを仏壇に供えます。もちろん宗派で若干違いはあります。

It seems like *Rokudo-mairi* was just yesterday.
Today is the day for the *Gozan no okuribi*, the bonfires on five mountainsides to guide the spirits of the deceased back to the spirit world. The *wagashi* shops of Kyoto are stocked with confections for Obon.
Obon offerings vary by the day:

Twelfth: White *mochi*, *omukae dango* (welcome dumpling), *hasugashi* (lotus confection)
Thirteenth: White *mochi*, *omukae dango*, *ohagi*, *hasugashi*
Fourteenth: White *mochi*, *shiramushi* (white steamed glutinous rice), *ohagi*
Fifteenth: White *mochi*, *okuri dango* (farewell dumpling), *shiramushi*
Sixteenth: *Okuri dango*, white *mochi*
The offerings, which are placed on the household Buddhist altar, may vary according to the Buddhist sect with which the family is associated.

子育飴　Kosodateame

みなとや幽霊子育飴本舗「幽霊子育飴」 麦芽糖、ざらめ糖
Minatoya Yurei Kosodateame honpo “Yurei Kosodateame” *Bakugato, Zarameto

六道詣りの時に買うのが「幽霊飴」正式な名前は「幽霊子育飴」とってもやさしい味。
ちなみにこの飴は、年中売っています。

The confection I buy at *Rokudo mairi* season is *Yurei ame* (ghost candy), whose proper name is *Yurei kosodate ame* (ghost nurturing candy). Its flavor is very plain.
Actually, *Yurei ame* is sold all year long.

勾玉池の白睡蓮 White water lilies in Magatama Pond

梅宮大社の北神苑にある勾玉池、池の中ほどに勾玉の形に区切ってあります。さすがにそこに咲く睡蓮を見ると格別の感じがする。いまは、白い睡蓮が咲いています。なんとも清楚な感じがしますね。

Magatama Pond, in the northern part of the Umemiya Grand Shrine precincts, has the shape of a curved sacred jewel (*magatama*) carved in its center. Perhaps that is why the lotuses there are something special.

亀屋良長「ほたる」　きんとき、粒あん
Kameya Yoshinaga "Hotaru" *Kintoki, Tsubuan

明治の「廃仏毀釈（はいぶつきしゃく）」の時に京都の街角からお地蔵さんが消えそうになった時、清水寺が引き受けたのがこのお地蔵さんたち。

This Jizo has become almost disappeared from the streets of Kyoto when “Haibutsukishaku” of the Meiji era, that's Kiyomizu Temple, has received argument.

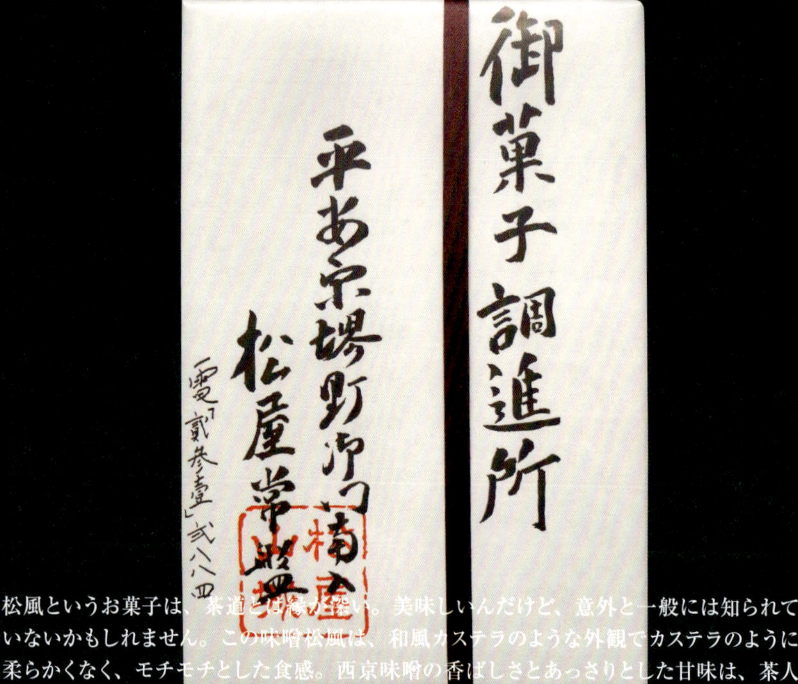

松風というお菓子は、茶道とは縁が深い。美味しいんだけど、意外と一般には知られていないかもしれません。この味噌松風は、和風カステラのような外観でカステラのように柔らかくなく、モチモチとした食感。西京味噌の香ばしさとあっさりとした甘味は、茶人好みです。このお店、創業承応年間なんだそうで、この味噌松風は、大徳寺の江月和尚の考案だそうです。 松屋常盤の「味噌松風」は、予約が必要。

松風という名前は、謡の『松風』に由来するそうです。
「浦寂し、鳴るは松風のみ」という一節を、裏に焼き色が付かないので寂しいということにかけた、いわば言葉遊びなんですって。

Matsukaze (breeze through the pines) is a confection with deep ties to the tea ceremony. It is delicious but, oddly enough, not widely known. *Miso Matsukaze* (*Matsukaze* made with miso, fermented bean paste) looks something like a castella sponge cake, but instead of being spongy, it is chewy, with the rich flavor of *Saikyo miso* and a crisp sweetness, very much to the taste of tea devotees. The shop that creates these sweets was founded, amazingly enough, in the 1650s, and the idea for *Miso Matsukaze* was suggested by Kogetsu, chief priest of the Daitokuji.

A reservation is required to purchase *Miso Matsukaze* at Matsuya Tokiwa.

The name is based on a bit of wordplay. “The lonely shore (*ura*), the only voice the wind in the pines” is a phrase from a Noh chant, *Matsukaze* (“the wind in the pines”). In the confection, only the underside (*ura*) is not toasted brown and, in its isolation, lonely; hence

長久堂「めぐみ」 外郎、備中白こしあん
Chokyudo "Megumi" *Uiro, Bicchu shirokoshian

鶴屋吉信「ほおづき」 外郎、白あん
Tsuruya Yoshinobu “Hozuki” *Uiro, Shiroan

この生菓子はリアルでしょ。
お盆といえばホオズキね。枝付きで精霊棚(盆棚)に飾り、死者の霊を導く提灯に見立てます。
このホオズキ、実を口の中にいれて口で鳴らす遊びがありましたね。

Isn’t this sweet incredibly realistic?
The *hozuki* (ground cherry or Chinese lantern plant) is a fixture at Obon. Decorating the altars set up at Obon, they symbolize the paper lanterns that guide the spirits of the dead back home. *Hozuki* are also fun to try to use as whistles.

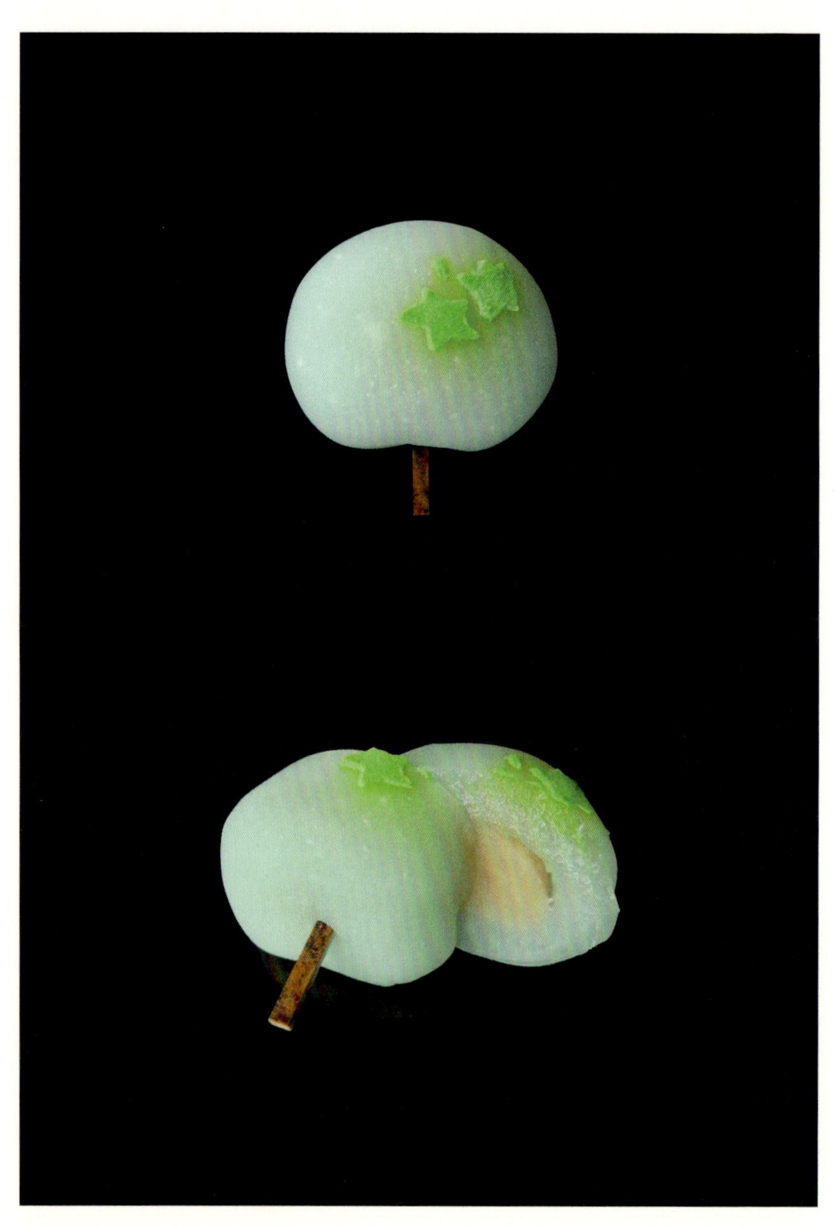

長久堂「涼 風」 外郎、備中白こしあん
Chokyudo "Suzukaze" *Uiro, Bicchu shirokoshian

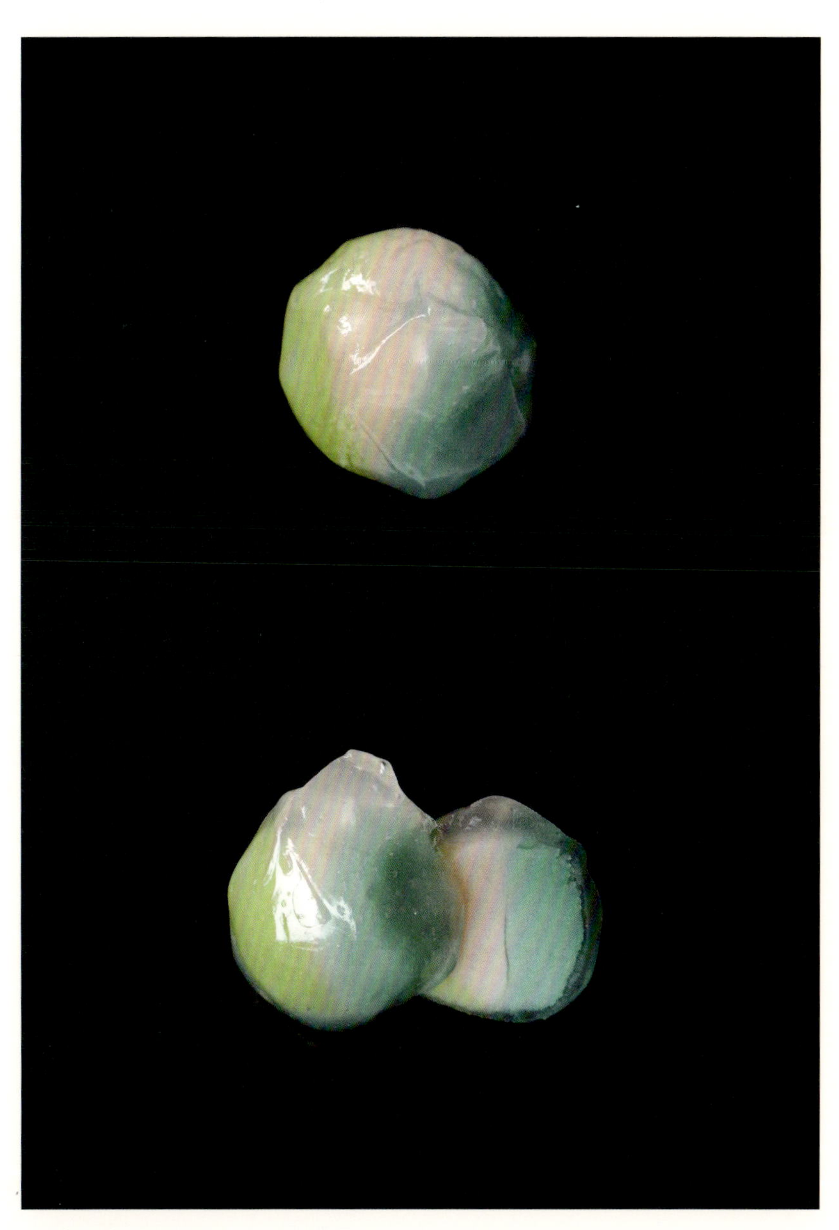

二條若狭屋「涼 風」 葛、白こしあん
Nijo Wakasaya "Suzukaze" *Kuzu, Shirokoshian

長久堂「京の朝」 外郎、備中白こしあん
Chokyudo "Kyo no Asa" *Uiro, Bicchu shirokoshian

この和菓子は、朝顔の蕾です。
This *wagashi* is inspired by a morning glory's bud.

塩芳軒「火の精」 葛（黒糖）、白こしあん
Shioyoshiken “Hi no Sei” *Kuzu with Kokuto, Shirokoshian

鶴屋吉信「宵待草」 焼皮、粒あん
Tsuruya Yoshinobu “Yoimachigusa” *Yakigawa, Tsubuan

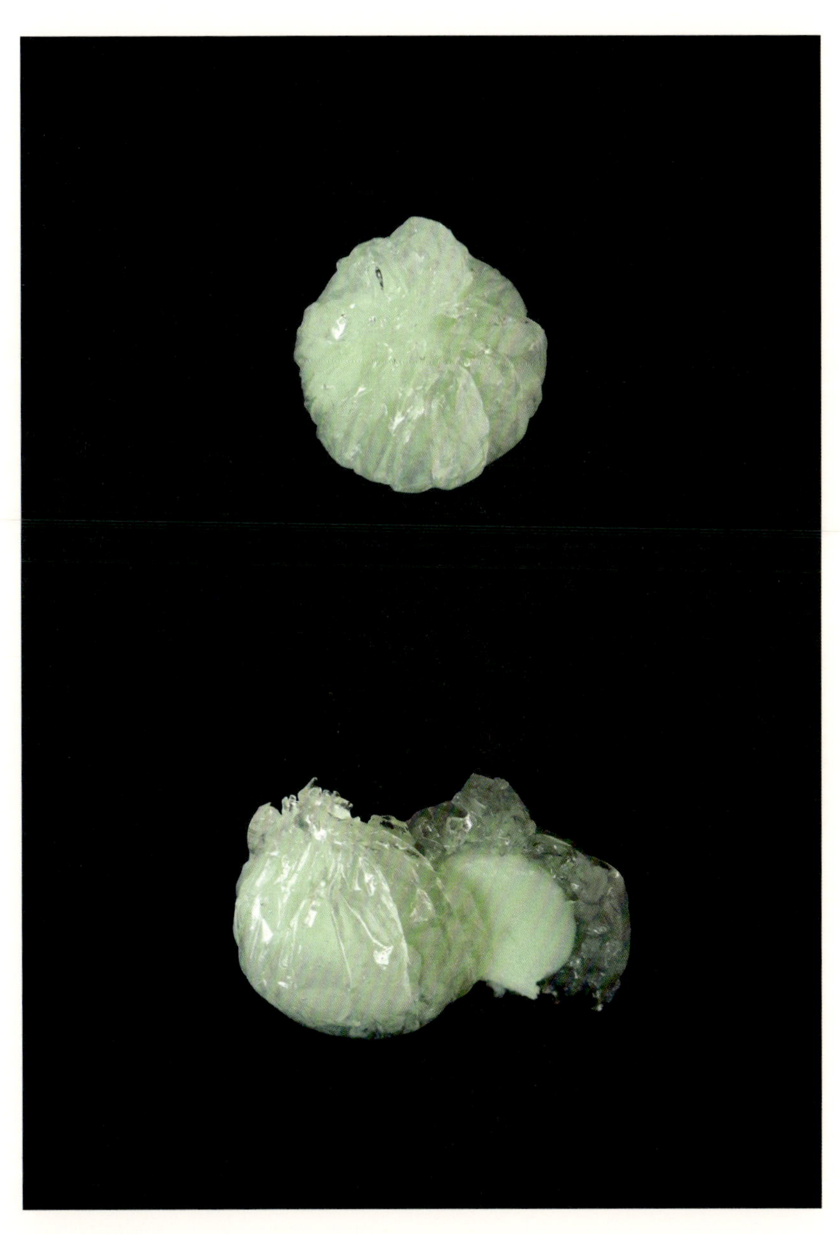

総本家駿河屋「さざ波」　錦玉、道明寺、白こしあん
Souhonke Surugaya “Sazanami (Wavelet)” *Kingyoku, Domyoji, Shirokoshian

奉納

地蔵盆は町内にあるお地蔵さんで行われます。
お地蔵さんは、子供たちを日頃から守ってくれています。
地蔵盆の当日は、お寺さんを呼んで読経をしてもらいます。
朝「数珠回し」をやる町内もあります。
直径二〜三メートルの大きな数珠を車座に座って読経にあわせて
順々にまわすというものです。
子供たちのためにお菓子を配ったり、福引きしたりもします。
福引きは「ふごおろし」といい、福引きでひいた景品を籠に入れて、
担当家の二階から紐で一階に下ろします。
こういうのを現在やっている町内は極めて少なくなりました。
市内の古くからある町内には子供がいなくなりましたから。

Jizobon is a festival celebrated at neighborhood Jizo shrines.

The bodhisattva Jizo is a protector of children.

On the day *Jizobon* is celebrated, a priest will be invited to chant the sutras.

Some neighborhoods also do a ceremony involving prayer beads, in the morning. Participants sit in a circle around a huge string of prayer beads, two or three meters across, and pass them around in time with the sutra chanting.

Snacks are served to the children, and they also get to participate in a drawing for prizes. The prizes are hung in a basket tied to a rope stretched from the second floors of a houses, and when someone wins a prize, it is lowered down to them.

Today, however, relatively few neighborhoods hold *Jizobon* festivals. Children are scarce in the older parts of the city.

鶴屋吉信「虫すだく」 上用、黒こしあん
Tsuruya Yoshinobu "Mushi Sudaku" *Joyo, Kurokoshian

autumn

長久堂「山 路」 外郎、備中白こしあん
Chokyudo “Yamaji” *Uiro, Bicchu shirokoshian

秋の色づき　Autumn Color

亀屋良長「菊 寿」 煉切り、白こしあん
Kameya Yoshinaga “Kikuju” *Nerikiri, Shirokoshian

二條若狭屋「菊 花」 きんとん（山芋）、黒粒あん
Nijo Wakasaya "Kikka" *Kinton with yam, Kurotsubuan

四条通沿いのみよしやが開いていて、しかも人が並んでいなかったので家族の土産にいつものように「タレ」と「きな粉」と2種みたらし団子を買う。竹の皮にいれてくれるのもいつもと同じ。
やっぱり賀茂みたらし茶屋みたいに、5個ついていてほしいな。ちなみに、どちらも離れている1個は頭、下の部分は胴体を表しています。頭に刺さっているつまようじがどうも怖いんですよ。
もうひとつ団子といえば、今宮神社のあぶり餅やね。

Miyoshiya on Shijo Avenue was open, without a line of customers waiting, so I popped in and bought some dumplings as a treat for my family. As usual, I chose two types, one with sweet sauce and one with powdery *kinako* (parched soybean powder). As usual, they wrapped them in bamboo peel.

I wish they put five dumplings on each skewer, as Kamo Mitarashi Chaya does. Here the dumpling at the end, separated from the others, stands for the head and the other rest for the body. The toothpicks stuck into the head are rather scary.

Another dumpling treat is the *aburimochi* (scorched dumplings) served at Imamiya Shrine.

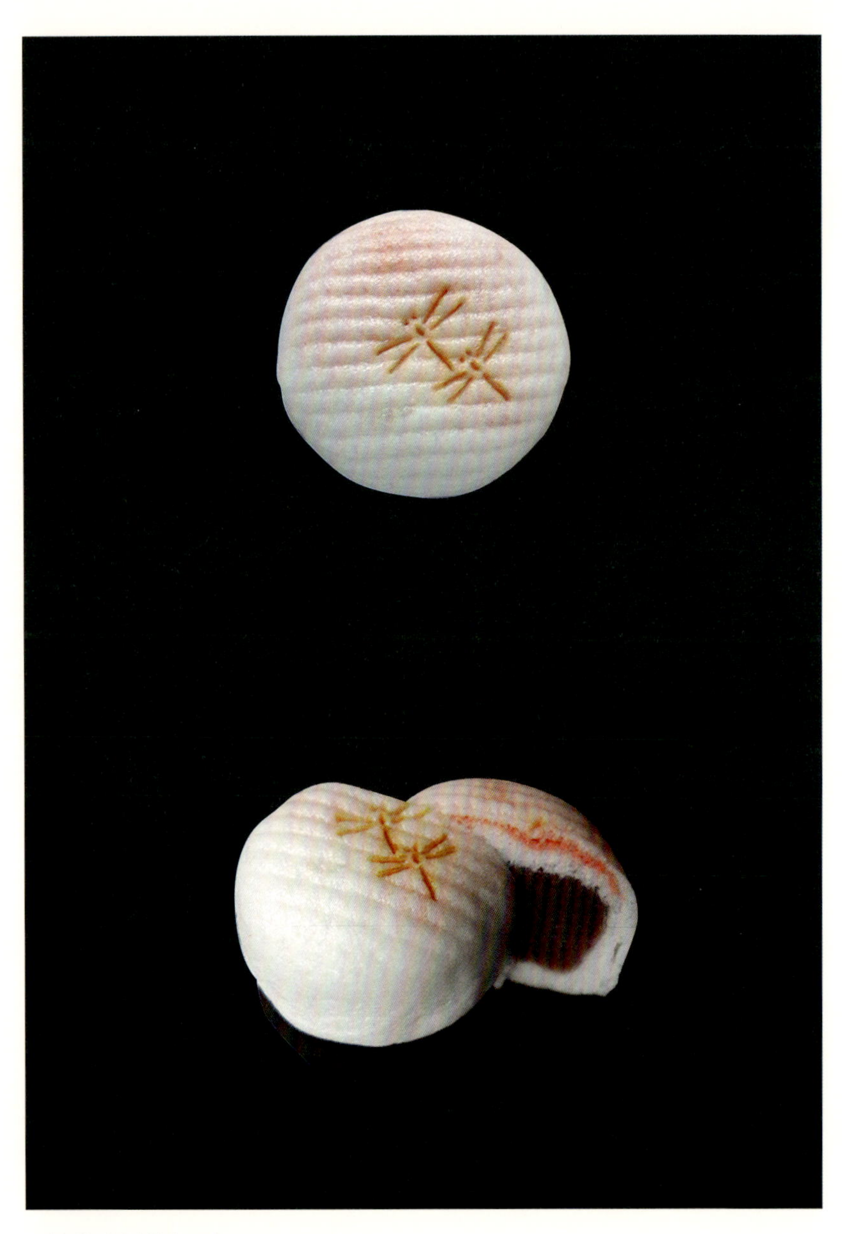

二條若狭屋「秋 日」 上用、こしあん
Nijo Wakasaya “Shujitsu” *Joyo, Koshian

もみじ　Autumnal tints

京都鶴屋鶴壽庵「栗粉餅」 きんとん、栗あん、黒粒あん
Kyoto Tsuruya Kakujuan “Kurikomochi” *Kinton, Kurian, Kurotsubuan

紫野源水「栗名月」 丹波栗、小豆こしあん
Murasakino Gensui “Kuri Meigetsu” *Tanba guri, Azuki koshian

萩の花　Hagi no Hana.

千本玉壽軒「こぼれ萩」 外郎、白こしあん

Senbon Tamajuken “Koborehagi” *Uiro, Shirokoshian

そろそろ萩の季節になってきました。先日、御所の東にある梨木神社に行ってきました。萩は万葉の時代には最も愛された秋草であり、その字もくさかんむりに秋を書いて表す、日本でできた国字です。咲く時期が秋のお彼岸の頃ということもあり、お彼岸につきものの「おはぎ」という言葉もこの萩からきている。粒あんの感じが萩の花に似ているからなんだそうです。

At last, the bush clover, *hagi*, are coming into bloom. The other day, I went to the Nashinoki Shrine, east of the imperial palace. The bush clover was the favorite autumn flower back in the days when the poems in the *Man'yoshu* were being composed. In fact, the character for *hagi*, which was devised in Japan, not borrowed from China, consists of the character for autumn under the grass radical: the iconic autumn plant itself. Because *hagi* bloom at the autumn equinox, their name has been given to the sweets served then. The unsieved bean paste on the outside is thought to resemble *hagi* flowers.

総本家駿河屋「友白髪」　煉切り、黒白こしあん
Souhonke Surugaya “Tomoshiraga” *Nerikiri, Kuro & shirokoshian

長久堂「年 祝（としのいわい）」 こなし、赤ごしあん
Chokyudo “Toshi no Iwai” *Konashi, Akagoshian

紫野源水「こぼれ萩」 きんとん、白小豆粒あん
Murasakino Gensui “Koborehagi” *Kinton, Shiroazuki tsubuan

千本玉壽軒「秋 桜」 きんとん、黒粒あん
Senbon Tamajuken "Akizakura" *Kinton, Kurotsubuan

長久堂「渡る風」　こなし、赤こしあん
Chokyudo “Wataru Kaze” *Konashi, Akakoshian

紫野源水「すすき野」 薯蕷、こしあん
Murasakino Gensui “Susukino” *Joyo, Koshian

すすき輝く　Japanese pampas grass shine in autumn

名月や・・・

名月や・・・

満 月　Full moon

長久堂「明月」 外郎、備中白あん
Chokyudo “Meigetsu” *Uiro, Bicchu shiroan

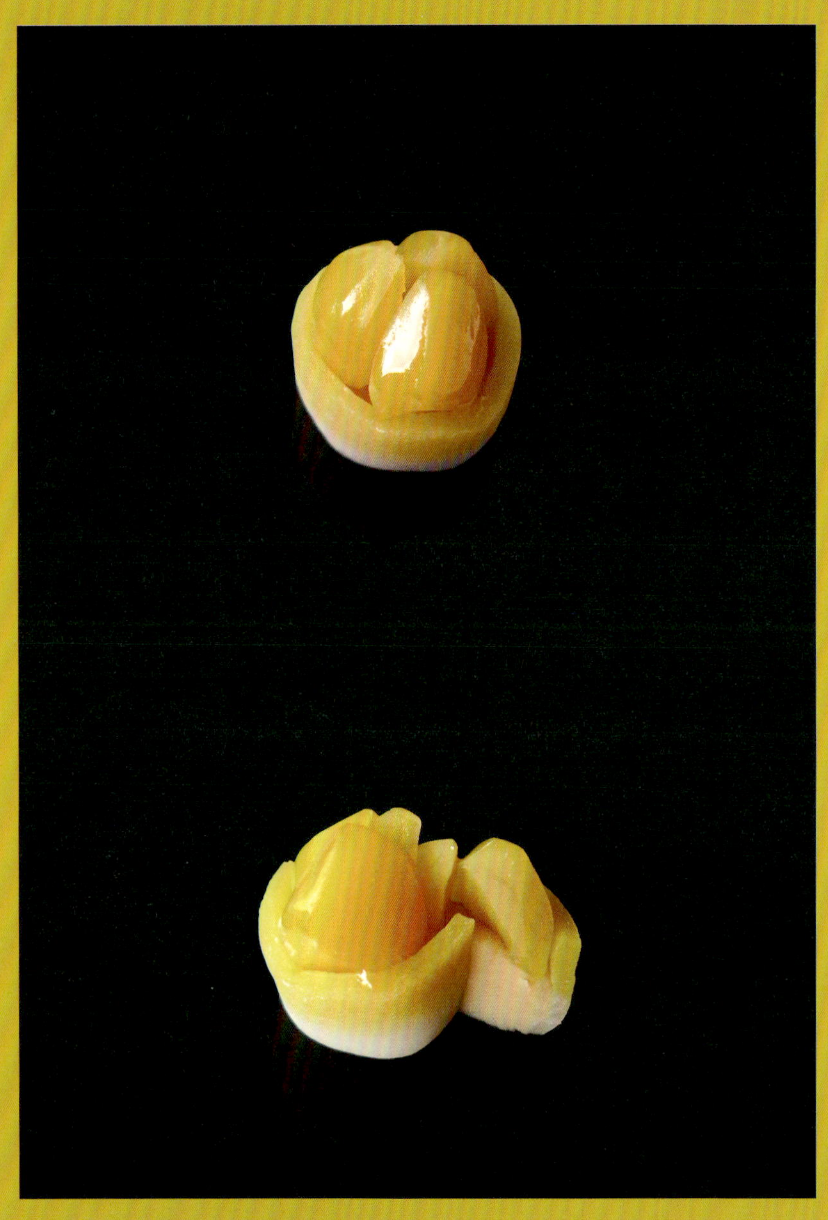

長久堂「栗名月」　外郎、懐中白こしあん
Chokyudo "Kuri Meigetsu" *Uiro, Kaichu shirokoshian

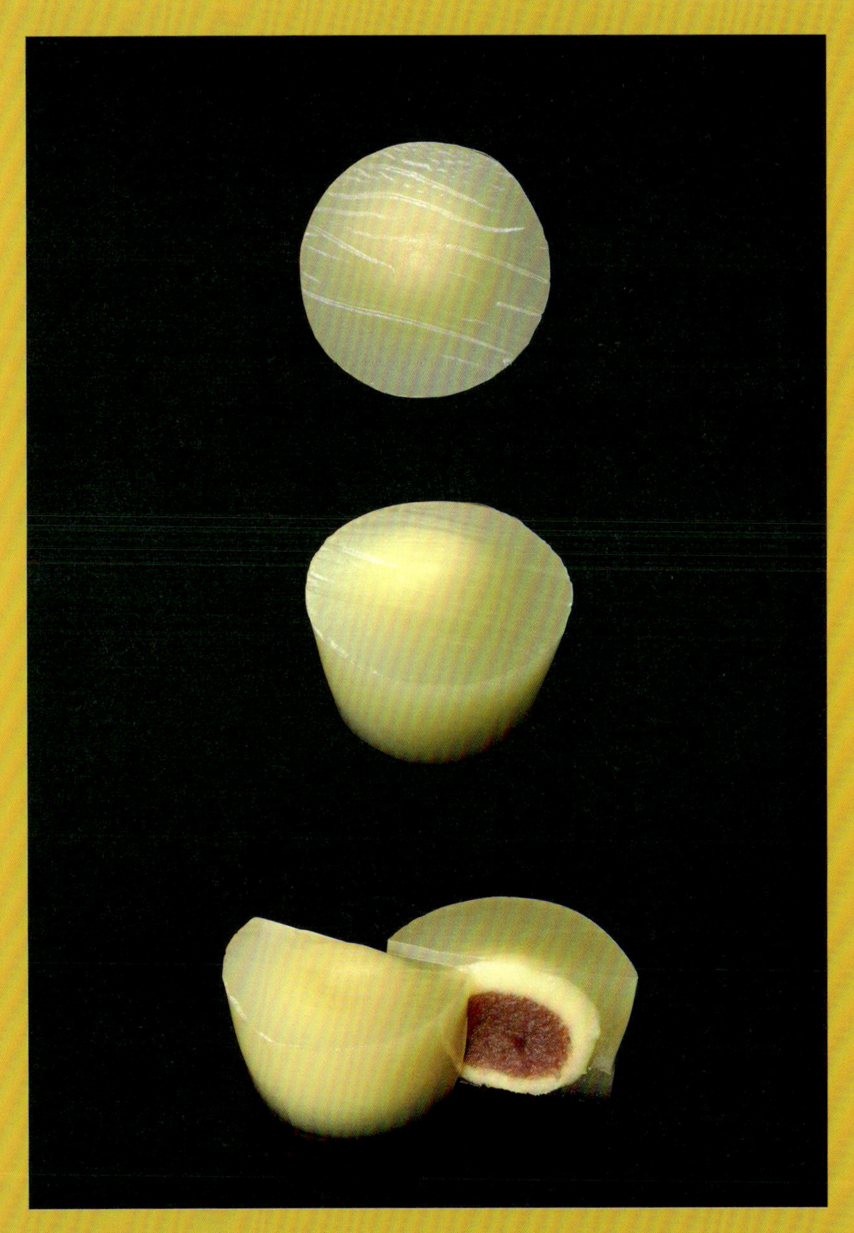

紫野源水「大沢の月」　半錦玉羹（煉切り、小豆あん）
Murasakino Gensui “Osawa no Tsuki” *Han-kingyokuan with Nerikiri, Azukian

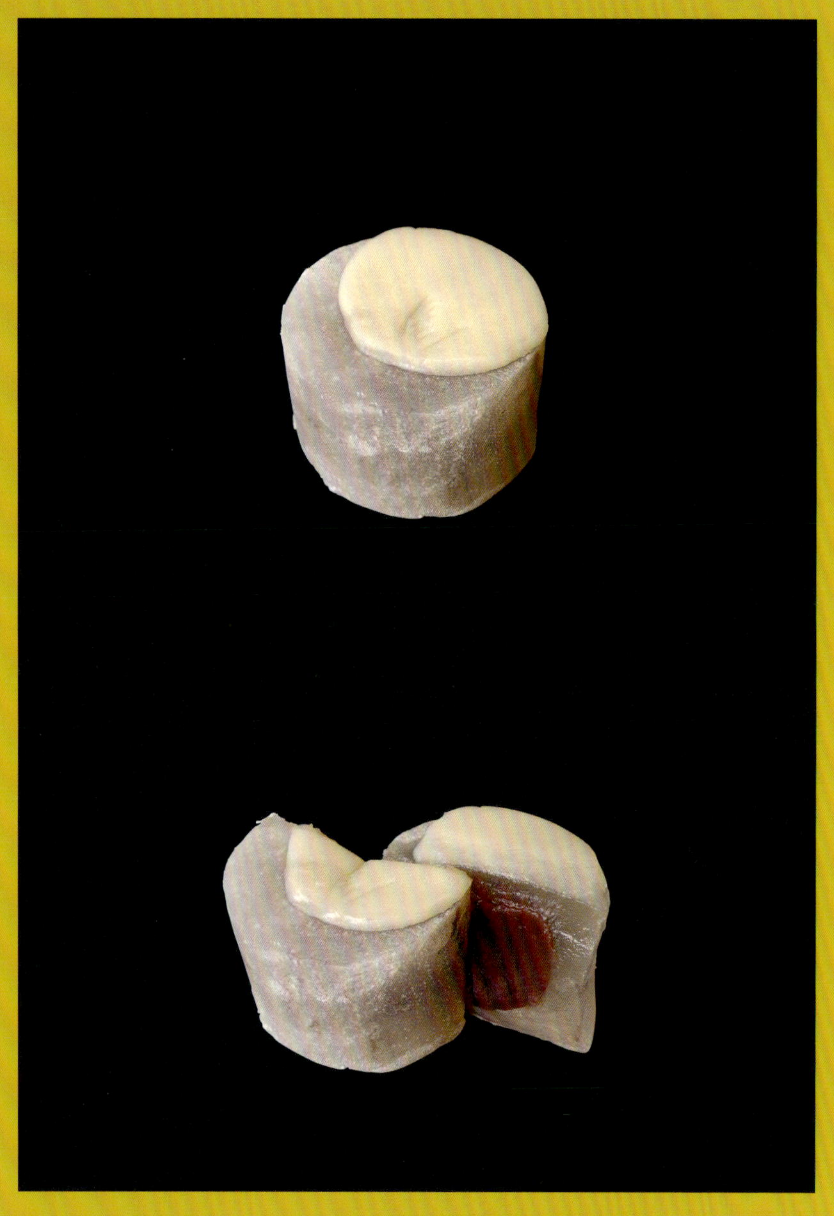

長久堂「月読みの道」　葛、こなし、赤こしあん
Chokyudo “Tsukuyomi no Michi” *Kuzu, Konashi, Akakoshian

塩芳軒「月見だんご」　外郎、黒こしあん
Shioyoshiken “Tsukimidango” *Uiro, Kurokoshian

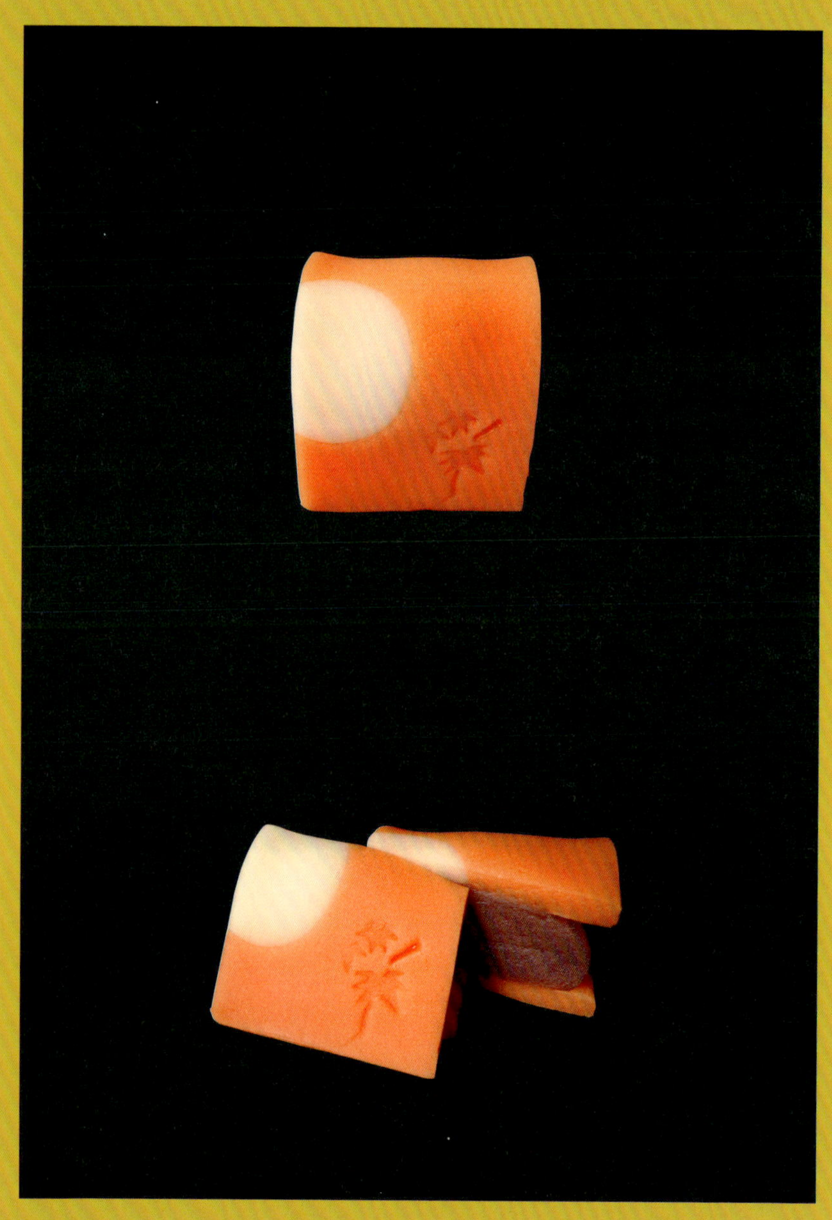

長久堂「名残の月」　こなし、赤こしあん
Chokyudo “Nagori no Tsuki” *Konashi, Akakoshian

鶴屋吉信「月 兎」 上用、黒こしあん
Tsuruya Yoshinobu “Tsukiusagi” *Joyo, Kurokoshian

二條若狭屋「芋名月」　こなし、黒こしあん
Nijo Wakasaya “Imo Meigetsu” *Konashi, Kurokoshian

二條若狭屋「芋名月」　こなし、こしあん
Nijo Wakasaya “Imo Meigetsu” *Konashi, Koshian

京都鶴屋鶴壽庵「おみなえし」 外郎、黒こしあん
Kyoto Tsuruya Kakujuan “Ominaeshi” *Uiro, Kurokoshian

総本家駿河屋「嵯峨菊」　煉切り、白こしあん
Souhonke Surugaya “Sagagiku” *Nerikiri, Shirokoshian

菊の着せ綿　Kiku no Kisewata

紫野源水「着せ綿」 煉切り（白小豆こしあん）
Murasakino Gensui “Kisewata” *Nerikiri with Shiroazuki koshian

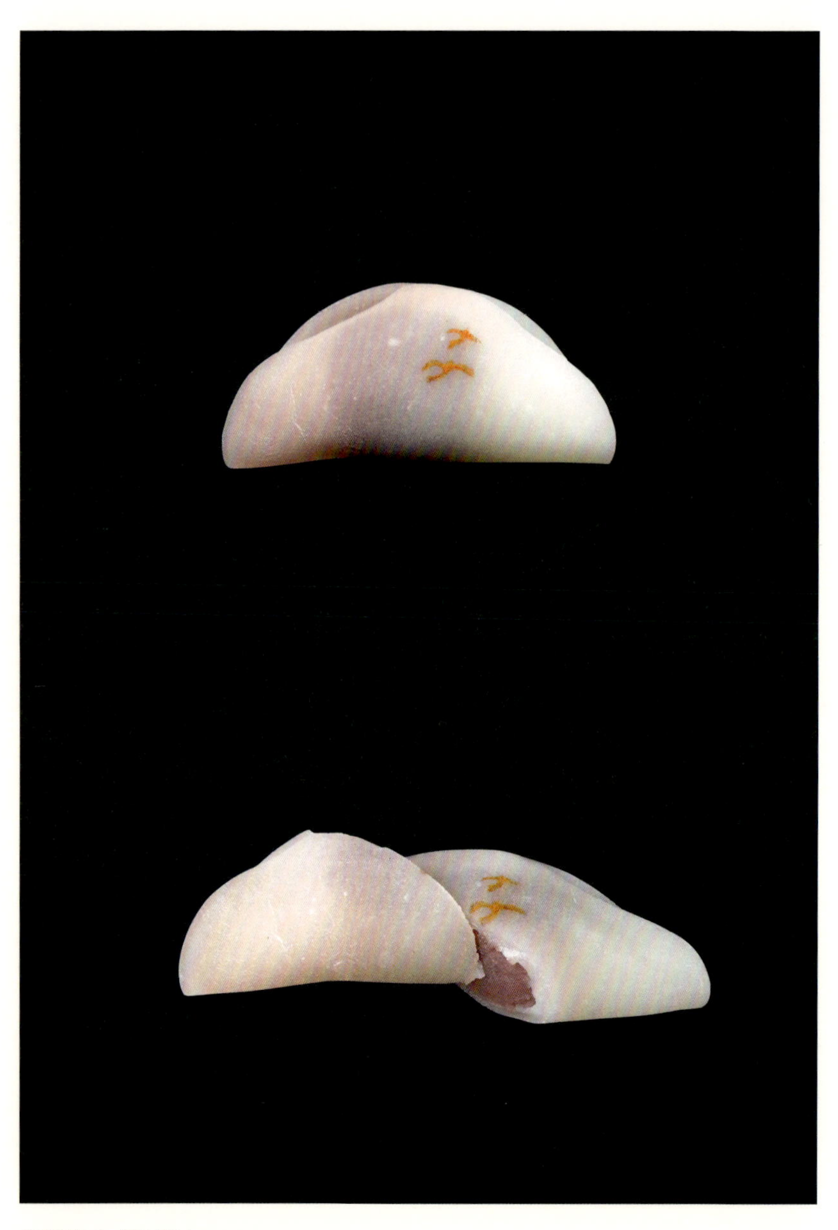

紫野源水「初 雁」 外郎、小豆こしあん入り

Murasakino Gensui “Hatsukari” *Uiro, Azuki koshian

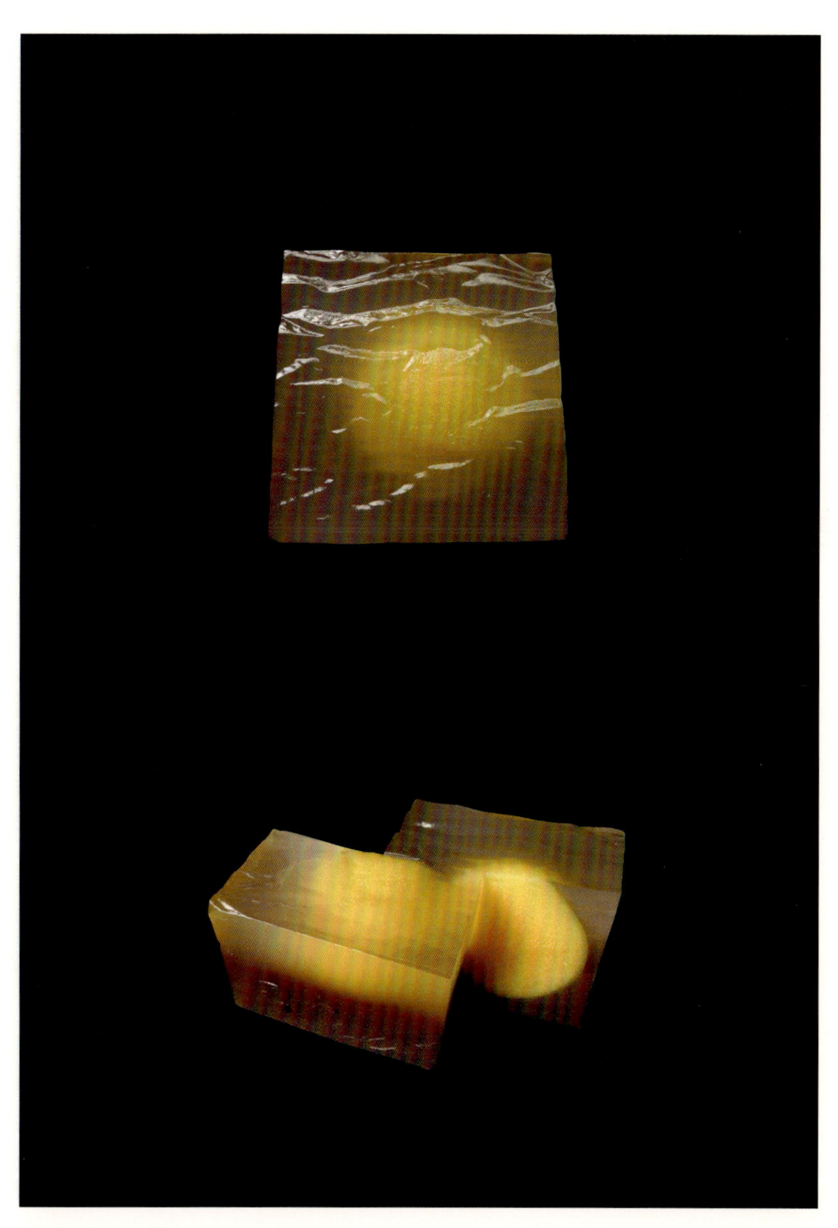

亀屋良長「湖中天」 葛羊羹、粒あん

Kameya Yoshinaga “Kochuten” *Kuzu yokan, Tsubuan

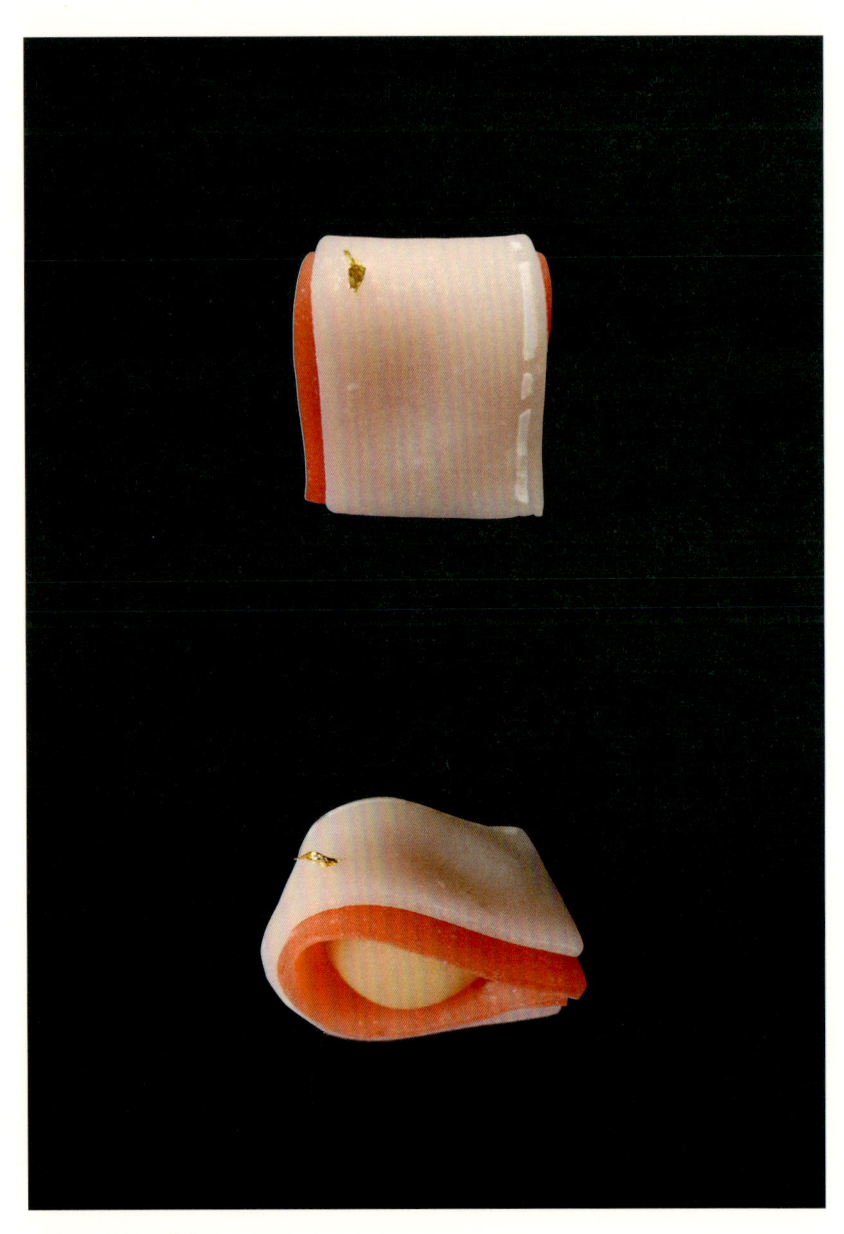

長久堂「京の雅」 外郎、備中白こしあん、金箔
Chokyudo "Kyo no Miyabi" *Uiro, Bicchu shirokoshian, Kinpaku

長久堂「栗拾い」 栗きんとん、赤ごしあん
Chokyudo "Kuri Hiroi" *Kuri kinton, Akagoshian

紅葉が咲く　Maples in flames

山水の庭　Japanese garden with rocks and sand

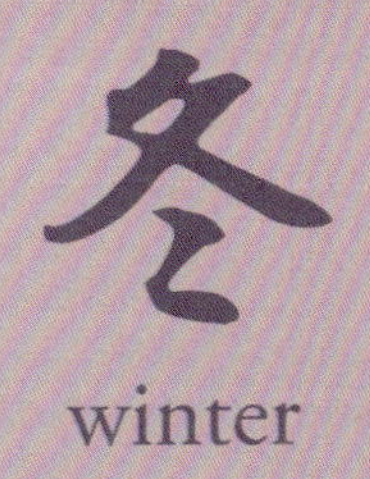
冬
winter

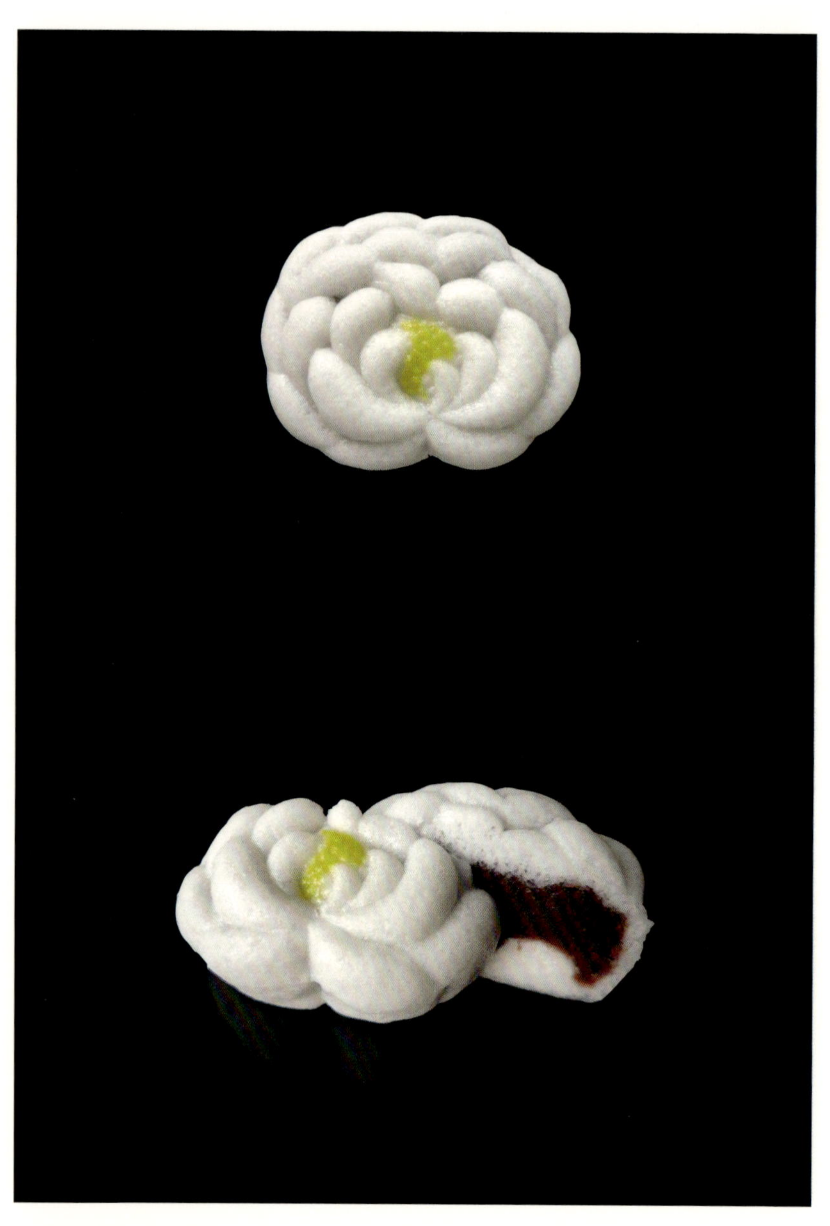

京都鶴屋鶴壽庵「菊上用」 上用、黒こしあん
Kyoto Tsuruya Kakujuan “Kikujoyo” *Joyo, Kurokoshian

献菊展　Dedicated the chrysanthemum

京都鶴屋鶴寿庵「延 年」 月餅、黒こしあん

Kyoto Tsuruya Kakujuan “Ennen” *Geppei, Kurokoshian

二條若狭屋「里の菊」 煉切り、白こしあん
Kameya Yoshinaga “Sato no Kiku” *Nerikiri, Shirokoshian

江戸時代の頃、男のたしなみとして「和菓子」というのがあったそうです。
・名前の由来の蘊蓄(うんちく)を語れる ・姿を愛でられる ・味を味わえる ・季節感を味わえる
といったことでしょうか。

Back in the Edo period, one of the accomplishments of the cultured man was expertise in *wagashi*. That entailed • Sharing a vast knowledge of the origins of *wagashi* names • Enjoying their appearance • Savoring their flavors • Appreciating their sense of the season

長久堂「秋の風」　こなし、赤こしあん
Chokyudo "Aki no Kaze" *Konashi, Akakoshian

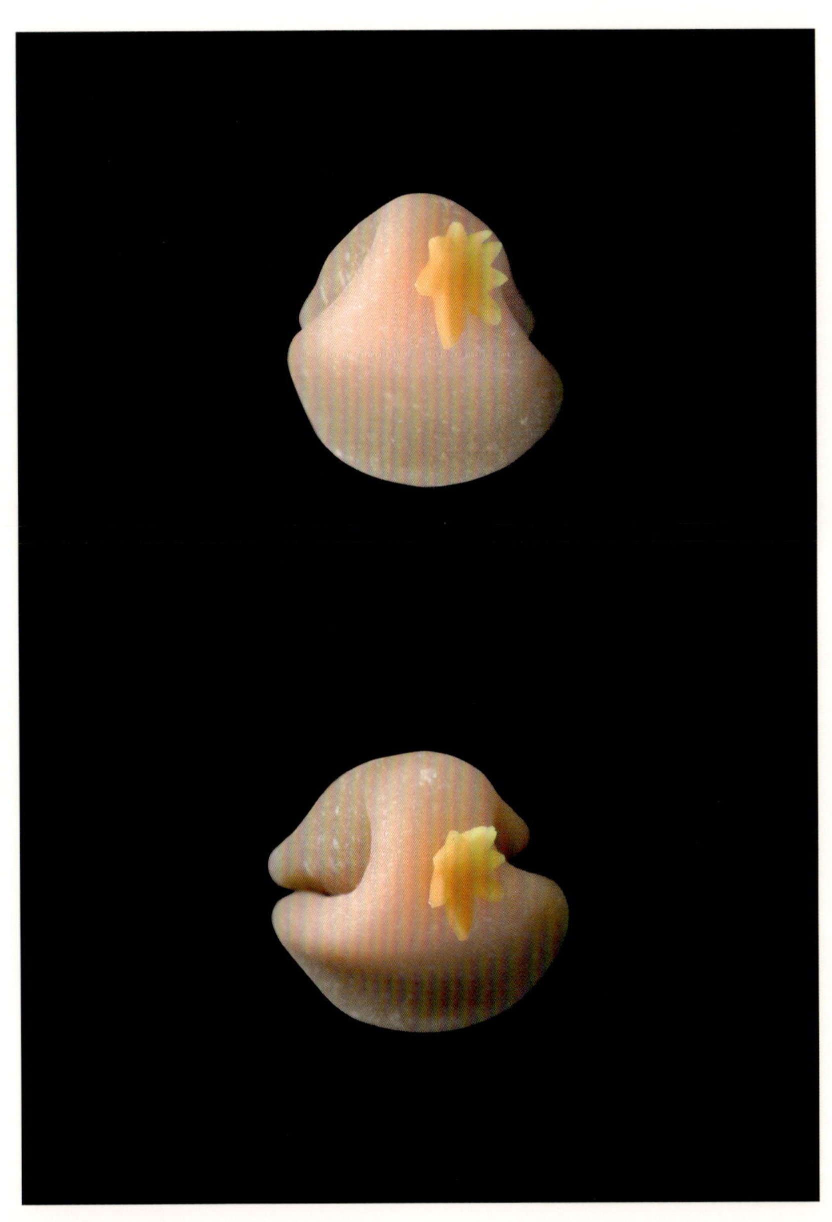

長久堂「芭蕉堂」 外郎、備中こしあん
Chokyudo "Bashodo" *Uiro, Bicchu koshian

十三夜は、日本独特の風習。
「小麦の名月」と呼ぶ地方もあったそうです。
旧暦の9月13日の夜のお天気で、翌年の小麦の豊作、凶作を占う習慣からきています。
「十三夜に曇りなし」という言葉があるぐらい。すっきり見えることが多い。

The night of the thirteenth of the ninth lunar month: the customs associated with this night are unique to Japan. In some regions, it is also called the "wheat moon," because it was possible to divine whether the next year's wheat harvest would be good or bad from the weather that night.

"The thirteenth and not a cloud in the sky": a proper statement of a bright outlook.

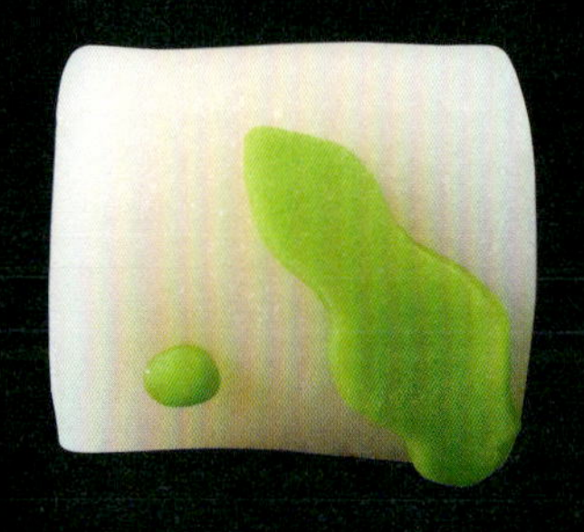

長久堂「后の月（あとのつき）」 外郎、煉切りあん
Chokyudo "Ato no Tsuki" *Uiro, Nerikirian

二條若狭屋「里の冬」 上用、黒こしあん
Nijo Wakasaya “Sato no Fuyu” *Joyo, Kurokoshian

二條若狭屋「亥の子餅」　外郎、黒つぶあん
Nijo Wakasaya "Inokomochi" *Uiro, Kurotsubuan

毎年、護王神社で「亥子祭（いのこまつり）」があります。
旧暦10月亥の日の夕方から翌朝の早朝にかけて行われます。

The Day of the Boar Festival begins at the Goou Shrine. It is held at the hour of the boar on the day of the boar in the month of the boar.

紫野源水「いのこ餅」 羽二重、こしあん

Murasakino Gensui "Inokomochi" *Habutae, Koshian

京都では、11月いっぱいまで亥の子餅を売っています。お火焚き饅頭同様、
収穫祭も兼ねているんでしょう。

Day of the boar mochi is available throughout November. Come to think of it, like the Ohitaki manju, that celebration doubles as a harvest festival.

京都鶴屋鶴壽庵「御所の秋」 きんとん、黒粒あん
Kyoto Tsuruya Kakujuan “Gosho no Aki” *Kinton, Kurotsubuan

そろそろ御所の銀杏とかも見頃かも……。
The gingko trees' foliage is probably aglow in the imperial palace grounds.

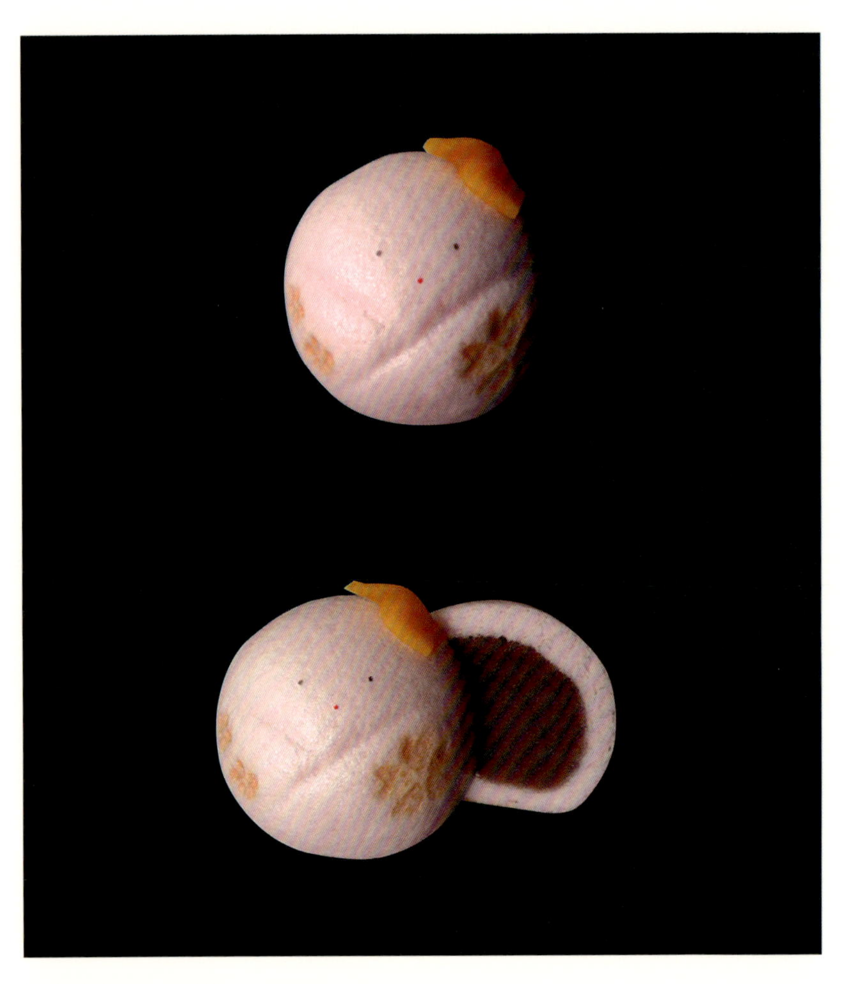

長久堂「およろこび」 上用、赤こしあん
Chokyudo "Oyorokobi" *Joyo, Akakoshian

11月15日は、七五三の日ですね。
この行事は、関東方面の年中行事で、本来京都ではやらんかったそうです。
京都は、十三詣りのほうが優先しました。七五三をするようになったのは比較的最近の話。

Today, the fifteenth of November, is Seven-Five-Three, when five-year-old boys and three- and seven-year-old girls make shrine visits to receive blessings for growth and happiness.

The custom is native to the Tokyo area, however, and was not performed Kyoto, where the shrine visit at age thirteen had priority. Only recently has Seven-Five-Three been adopted here.

二條若狭屋「初 雪」 黒糖きんとん、黒つぶあん
Nijo Wakasaya “Hatsuyuki” *Kokuto kinton, Kurotsubuan

赤く変わるのが「紅葉(こうよう)」
黄色に変わるのを「黄葉（こうよう、おうよう）」
褐色に変わるのを「褐葉(かつよう)」と呼ぶそうです。
万葉集の歌に詠まれる「もみじ」は、「黄葉」と書かれているものが圧倒的で、
「紅葉」はごくわずかなんだそうです。

Terms for autumn foliage vary according to the color of the leaves — are they red, yellow, or brown?

While red foliage is now synonymous with autumn, in the poems in the *Man'yoshu*, yellow foliage was referred to far more often, and red foliage very rarely. Times have changed.

紫野源水「紅 葉」 煉切り、白小豆こしあん
Murasakino Gensui "Koyo" *Nerikiri, Shiroazuki koshian

紫野源水「織部薯蕷」 こしあん

Murasakino Gensui “Oribe Joyo” *Koshian

いろんな生菓子があるけれど、結局こういう薯蕷（上用）饅頭が飽きがこないというか。
職人さんの腕がわかるのはこういうものですね。

Of the myriad varieties of moist sweets, these joyo manju (steamed buns made with yams) are a constant favorite. They clearly display the skill of the confectioner.

紫野源水「錦 秋」 きんとん、つぶあん
Murasakino Gensui “Kinshu” *Kinton, Tsubuan

紫野源水「寒 菊」 外郎、白紅あん
Murasakino Gensui “Kangiku” *Uiro, Shirobenian

嵯峨菊　Sagagiku

長久堂「錦 秋」　こなし、備中白こしあん
Chokyudo “Kinshu” *Konashi, Bicchu shirokoshian

塩芳軒「いちょう」　蕨羽二重、黒粒あん
Shioyoshiken “Icho” *Warabi habutae, Kurotsubuan

二條若狭屋「もみじ狩り」 羊羹、栗あん
Nijo Wakasaya “Momijigari” *Yokan, Kurian

千本釈迦堂の「大根炊き」 Daiko daki (Boiled White raddish)

いちめんの紅葉 Red carpet

亀屋良長「大きなかぶ」 上用、黒粒あん
Kameya Yoshinaga “Okina Kabu” *Joyo, Kurotsubuan

紫野源水「木枯らし」　そば薯蕷、つぶあん

Murasakino Gensui “Kogarashi” *Soba joyo, Tsubuan

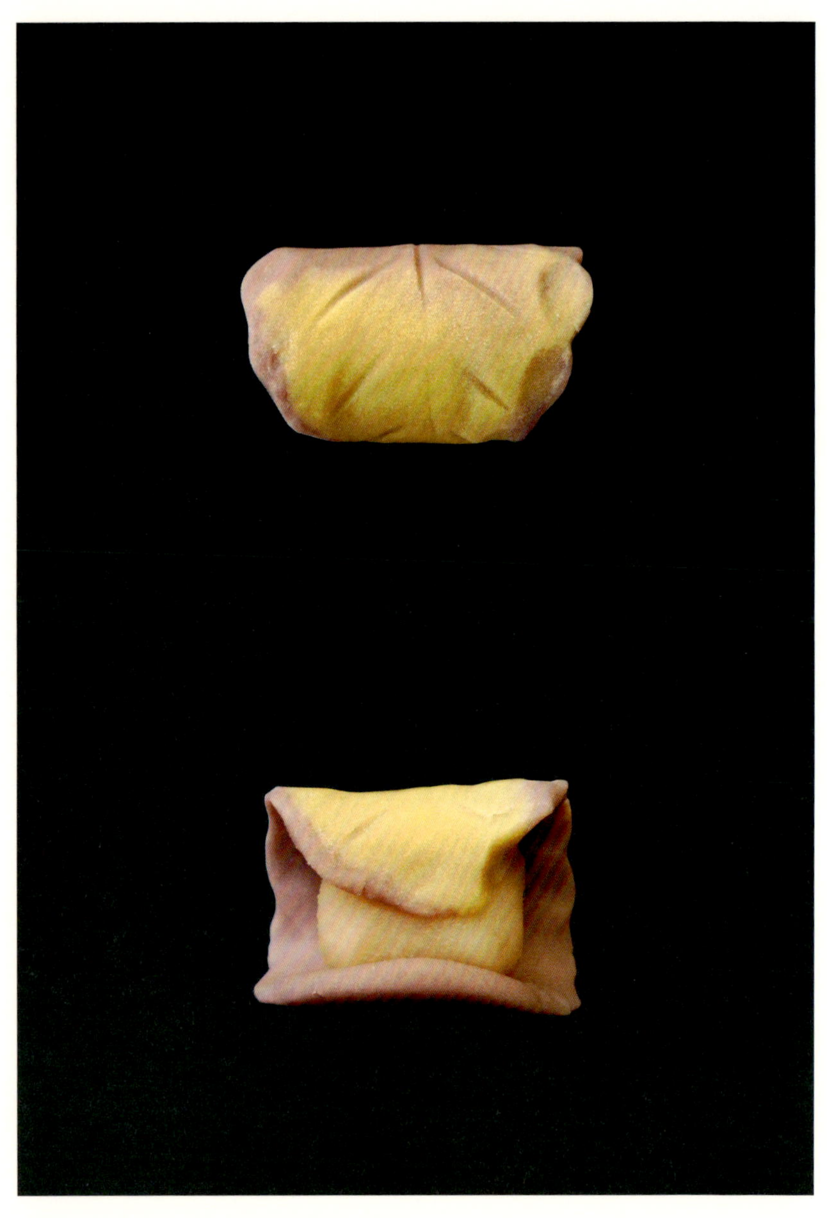

紫野源水「冬木立」 煉切り、白小豆こしあん
Murasakino Gensui “Fuyukodachi” *Nerikiri, Shiroazuki koshian

二條若狭屋「枯 葉」 外郎、黒粒あん
Nijo Wakasaya “Kareha” *Uiro, Kurotsubuan

冬木立 Winter grove

紫式部　Murasakishikibu

椿もいろんな種類が咲いています。
いままで紅葉で目立たなかっただけかもしれませんが椿もきれいで、
日本の花という感じがします。

Several varieties of camellias have begun blooming.

They may not have stood out among the autumn foliage thus far, but the camellias are lovely. To me, they seem the very flower of Japan.

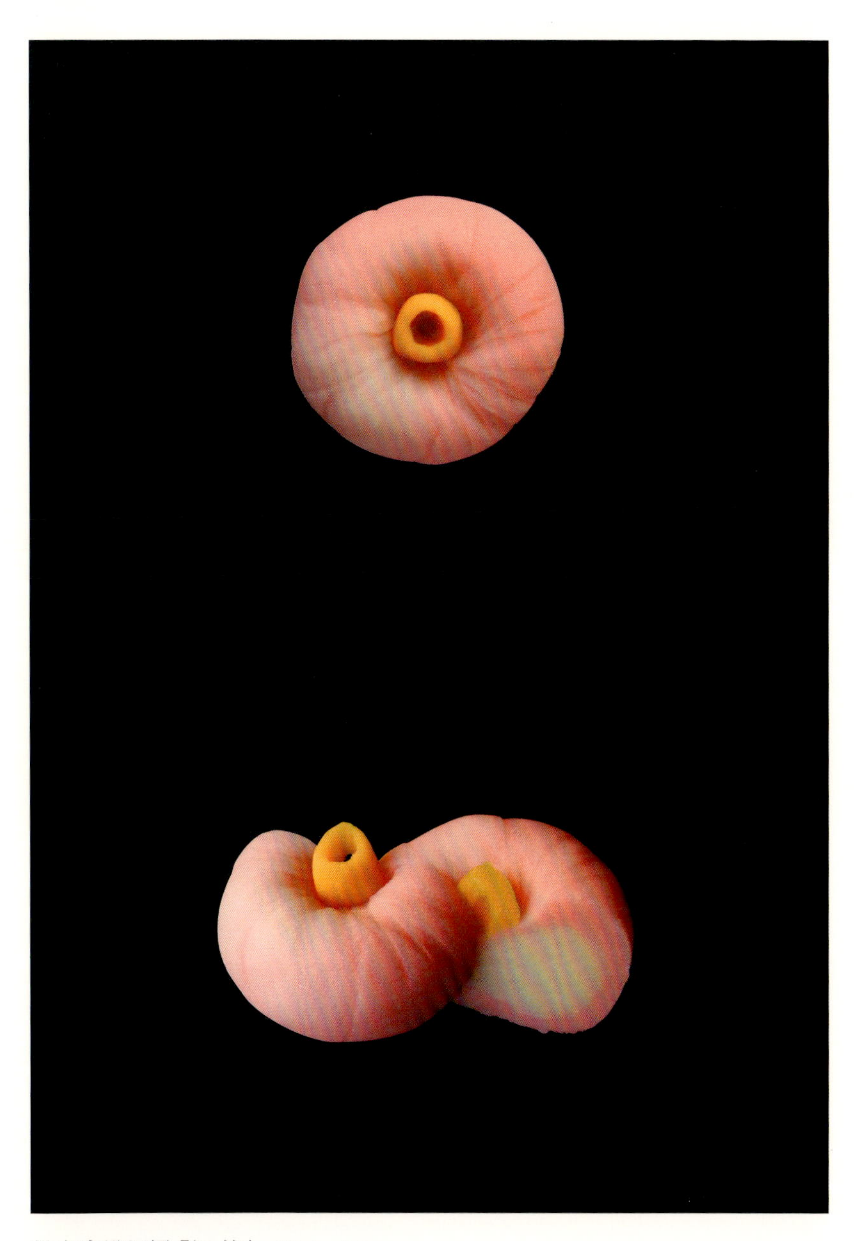

総本家駿河屋「紅 椿」 煉切り、白あん
Souhonke Surugaya “Benitsubaki” *Nerikiri, Shiroan

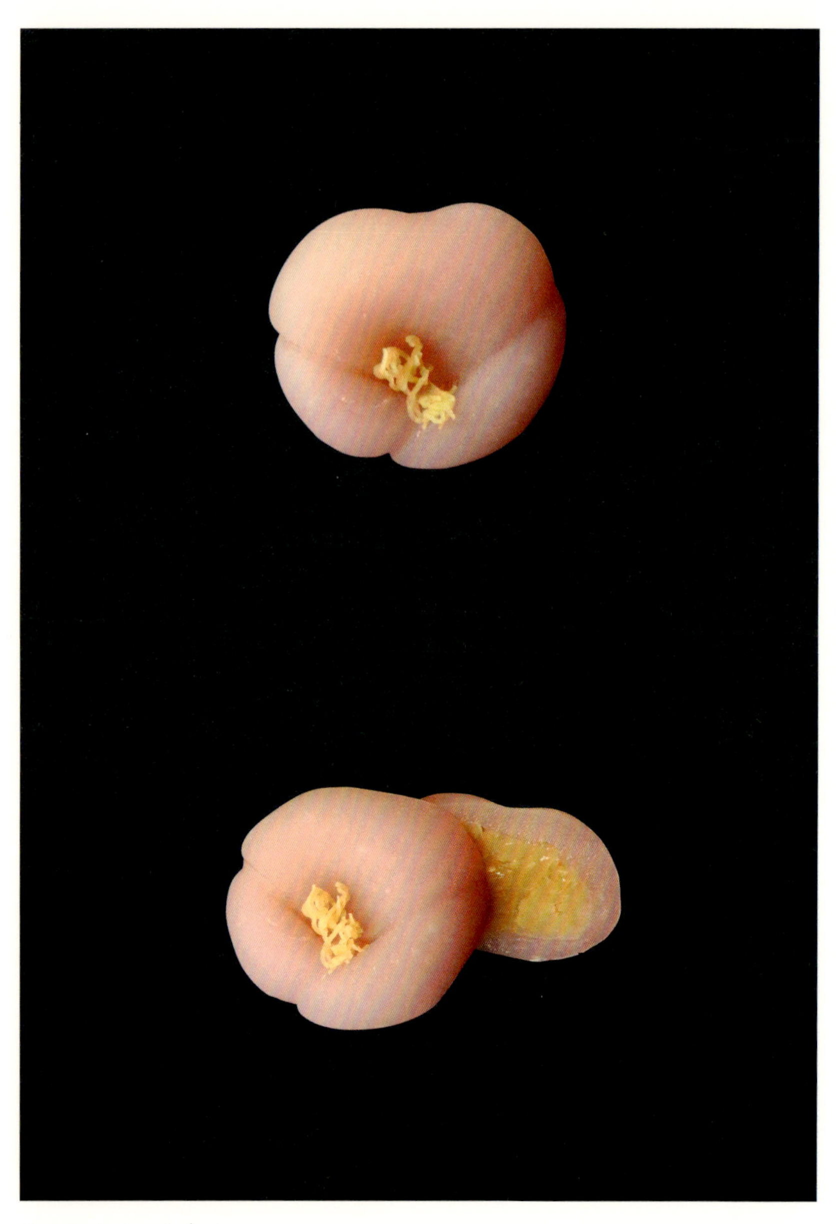

紫野源水「さざんか」 外郎、白小豆粒あん
Murasakino Gensui “Sazanka” *Uiro, Shiroazuki tsubuan

紫野源水「初 雪」 きんとん、白小豆粒あん
Murasakino Gensui "Hatsuyuki" *Kinton, Shiroazuki tsubuan

長久堂「すこやかに」 こなし、赤ごしあん
Chokyudo "Sukoyakani" *Konashi, Akagoshian

京都の伝統野菜「鹿ケ谷かぼちゃ」
The *shishigatani* squash is one of the vegetables traditionally raised in Kyoto.

京都鶴屋鶴壽庵「冬小立」　白月餅、黒あん
Kyoto Tsuruya Kakujuan “Fuyukodachi” *Shiro geppei, Kuroan

総本家駿河屋「万 両」 煉切り、黒こしあん
Souhonke Surugaya “Manryo” *Nerikiri, Kurokoshian

冬牡丹または寒牡丹　Winter Peony or Peony bloom in winter

普通の牡丹と同じ品種ですが、雪よけのワラ囲いなど特別な管理をして、
早春に咲かせるようにしたものです。
「そのあたり ほのとぬくしや 寒ぼたん」高浜虚子
Tree peony flowers with care, they can bloom in the winter.
“A winter peony, yes, but with a hint of warmth”— Takahama Kyoshi

二條若狭屋「寒牡丹」　煉切り、黒こしあん
Nijo Wakasaya “Kanbotan” *Nerikiri, Kurokoshian

この寒牡丹の種類は寒豊明。
This tree peony varietal is called *Kanhomei* .

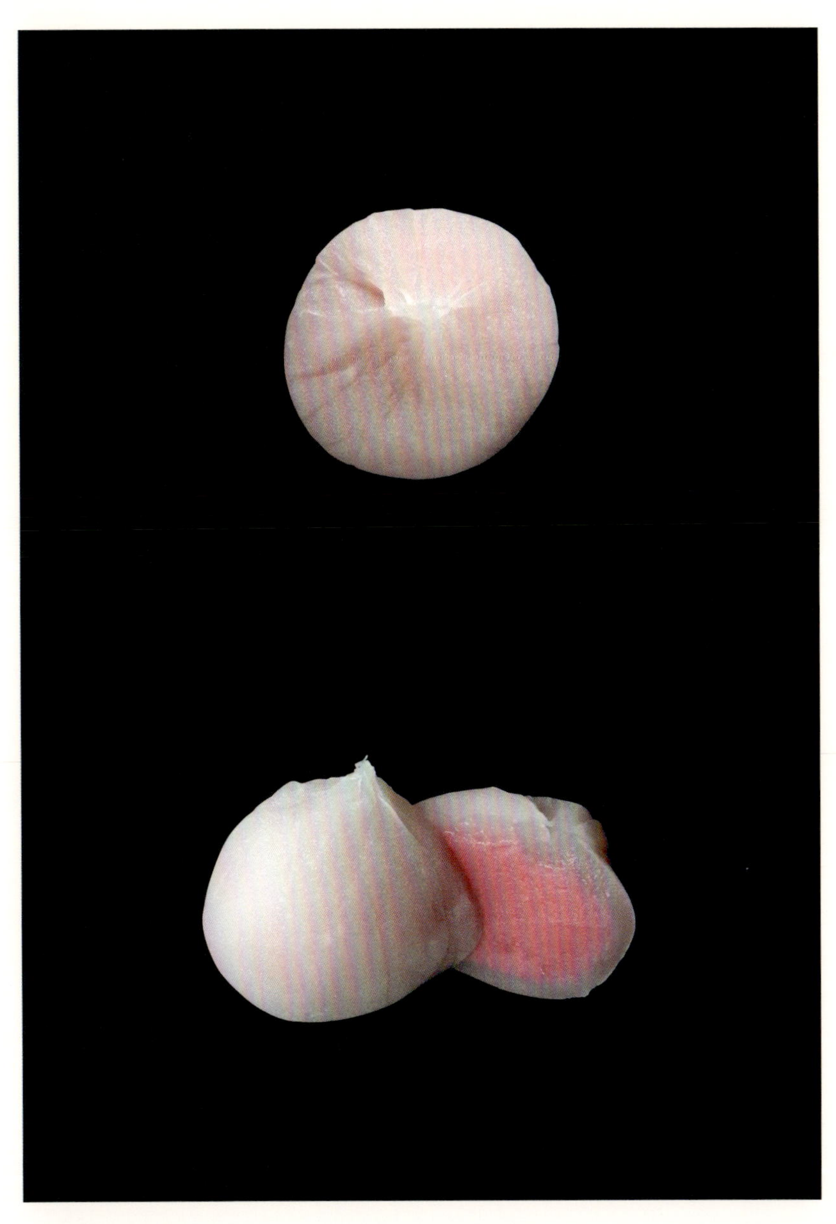

京都鶴屋鶴壽庵「冬牡丹」　白月餅、薄紅あん
Kyoto Tsuruya Kakujuan “Fuyubotan” *Shiro geppei, Usubenian

紫野源水「寒牡丹」 煉切り、白小豆こしあん
Murasakino Gensui “Kanbotan” *Nerikiri, Shiroazuki koshian

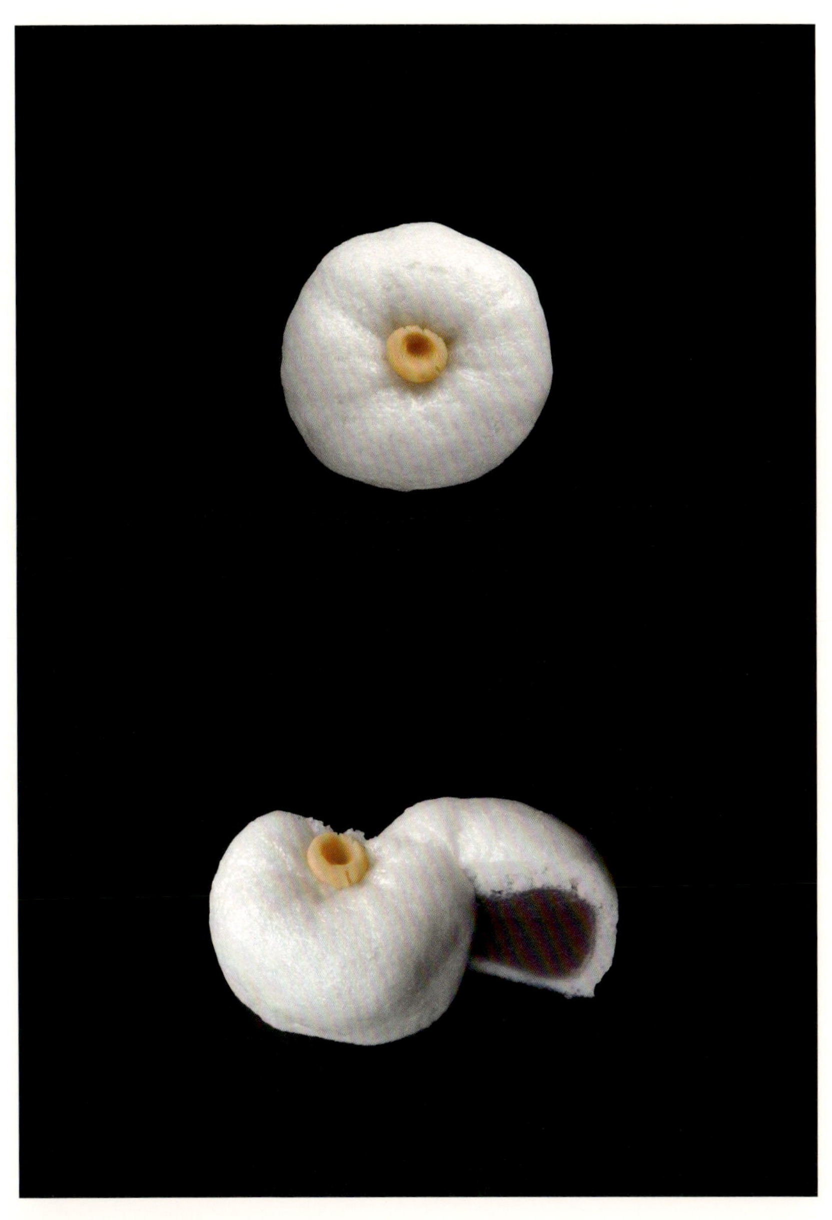

紫野源水「白玉椿」 薯蕷、小豆こしあん
Murasakino Gensui “Shirotama Tsubaki” *Joyo, Azuki koshian

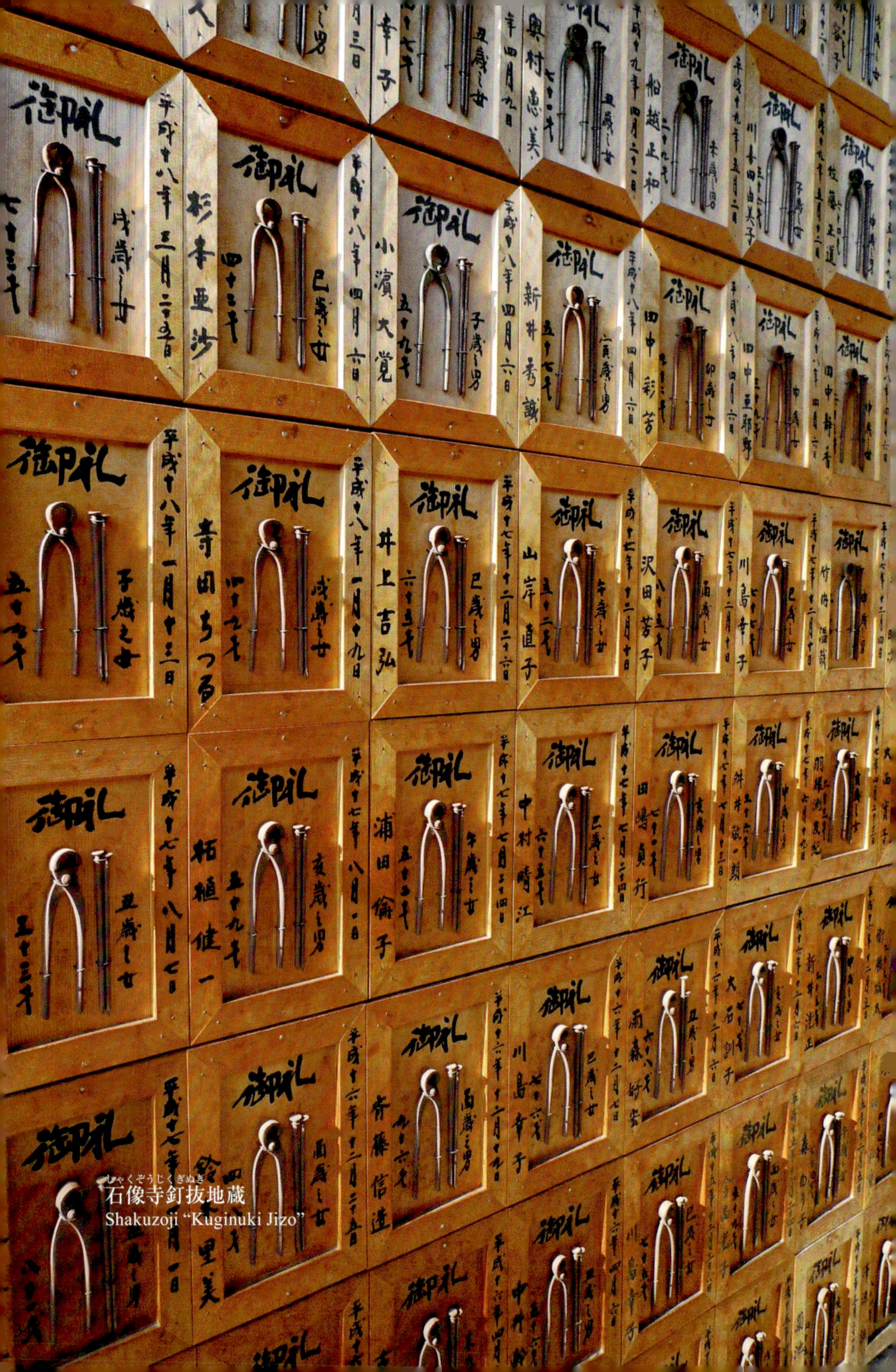

石像寺釘抜地蔵（しゃくぞうじくぎぬきじぞう）
Shakuzoji “Kuginuki Jizo”

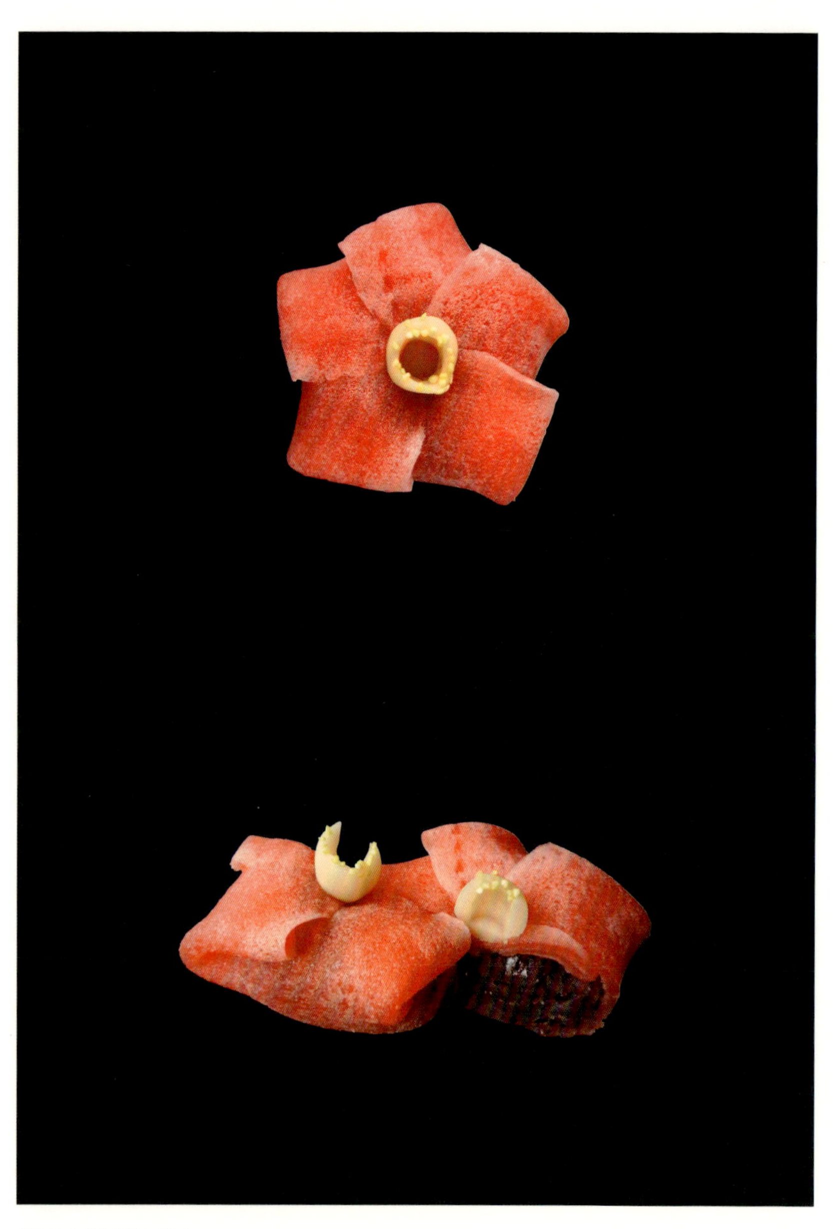

鶴屋吉信「寒 椿」 焼皮、粒あん
Tsuruya Yoshinobu “Kantsubaki” *Yakigawa, Tsubuan

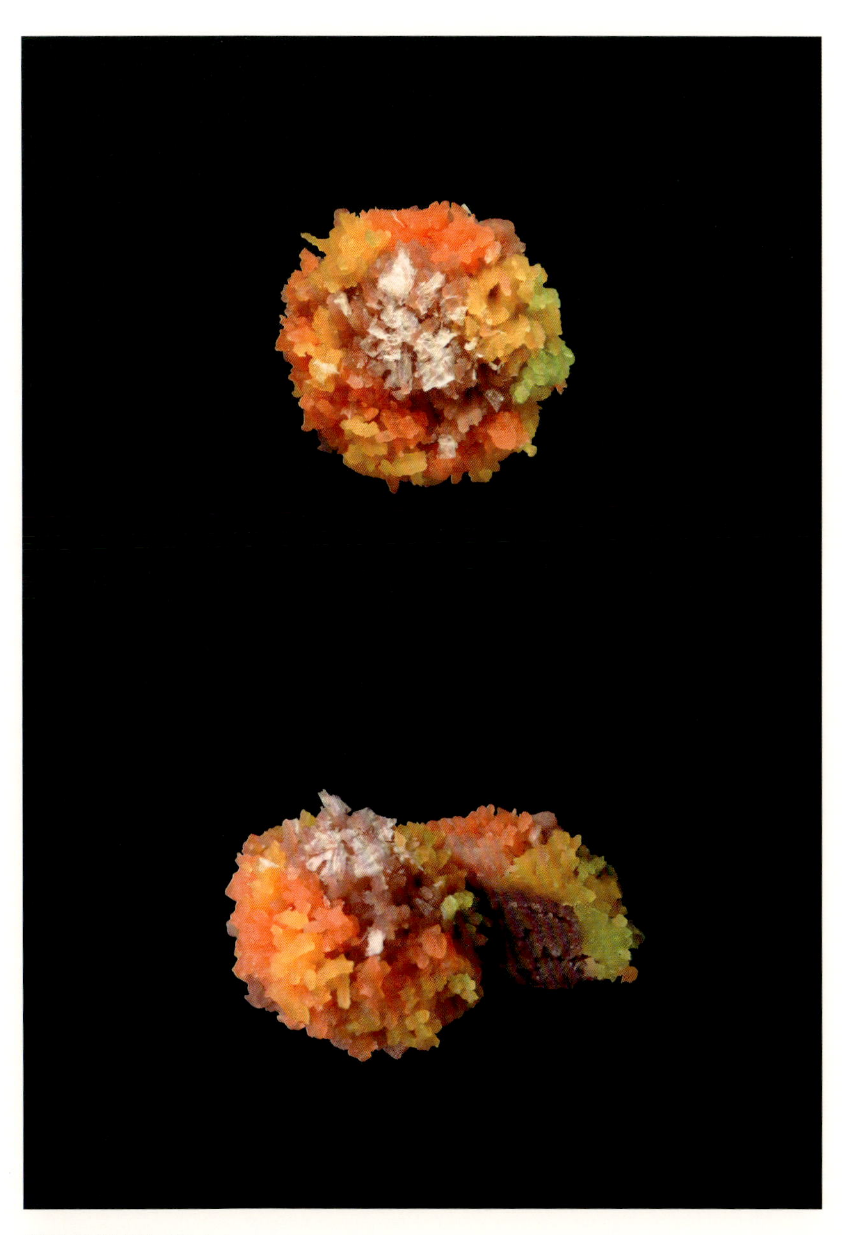

鶴屋吉信「冬ごこち」 きんとん、粒あん
Tsuruya Yoshinobu “Fuyugokochi” *Kinton, Tsubuan

亀屋良長「藪柑子」 煉切り、黒こしあん
Kameya Yoshinaga “Yabukoji” *Nerikiri, Kurokoshian

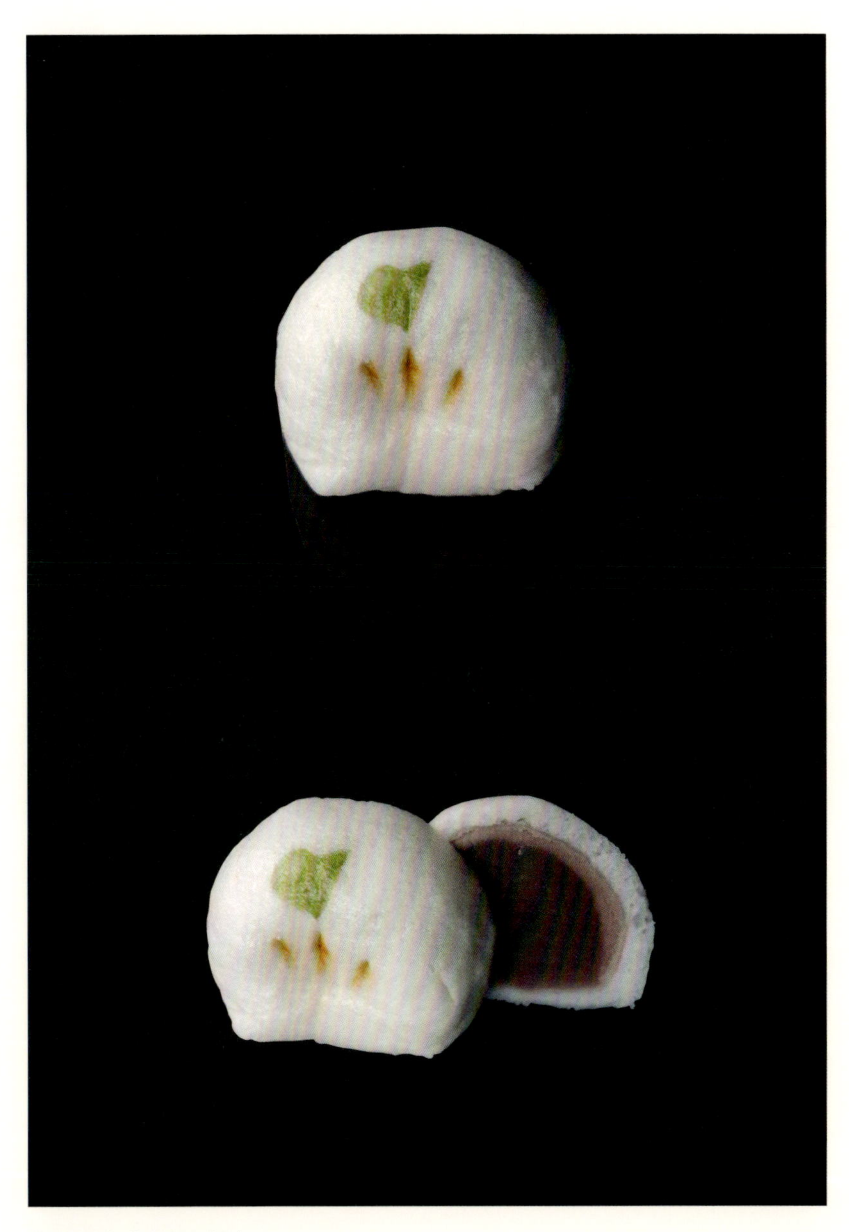

紫野源水「雪中の松」 薯蕷、こしあん
Murasakino Gensui “Secchu no Matsu” *Joyo, Koshian

亀屋良長「切 梅」 煉切り、黒こしあん
Kameya Yoshinaga “Kiriume” *Nerikiri, Kurokoshian

長久堂「冬の紅」 きんとん、粒あん
Chokyudo “Fuyu no Kurenai” *Kinton, Tsubuan

本書に掲載しております和菓子はほとんどが上菓子、生菓子と言われるものです。
これらのお菓子はひとつひとつ手作りのため、全く同じものは出来ません。
また季節や天候によっても出来上がりが左右されるものが多く、それぞれの菓子司に
ご注文いただいても、同じものが手に入るとは限りません。
お菓子との出会いも楽しみとご理解ください。
本書に御掲載をお願いいたしました、和菓子店の住所、こちらに記載させていただきました。

和菓子司

鍵善良房
京都市東山区祇園町北側 264 番地
電話 (075) 561-1818　FAX (075) 525-1818　http://www.kagizen.co.jp

亀屋清永
京都市東山区祇園石段下南 534
電話 (075) 561-2181　FAX (075) 541-1034

亀屋良長　亀屋良長株式会社
京都市下京区四条通油小路西入柏屋町 17-19
電話 (075) 221-2005　FAX (075) 223-1125　http://kameya-yoshinaga.com

御菓子司　河藤
大阪市天王寺区四天王寺 1-9-21
電話 (06) 6771-6906

京都鶴屋鶴壽庵
京都市中京区壬生梛ノ宮町 24
電話 (075) 841-0751　FAX (075) 841-0707　http://www.kyototsuruya.co.jp

塩芳軒
京都市上京区黒門通中立売上ル
電話 (075) 441-0803

末富
京都市下京区松原通り室町東入ル
電話 (075) 351-0808　FAX (075) 351-8450

千本玉壽軒
京都市上京区千本通今出川上ル上善寺町 96
電話 (075) 461-0796　FAX (075) 464-6717

総本家駿河屋　伏見本舗
京都市伏見区京町 3 丁目 190
電話 (075) 611-5141　FAX (075) 611-5142　http://www.souhonke-surugaya.co.jp

長久堂
京都市北区上賀茂畔勝町 97-3
電話 (075) 712-4405　FAX (075) 712-3585

鶴屋吉信
京都市上京区今出川通堀川西入
電話 (075) 441-0105　FAX (075) 431-1234　http://www.turuya.co.jp

中村軒　株式会社 中村軒
京都市西京区桂浅原町 61
電話 (075) 381-2650　http://www.nakamuraken.co.jp

二條若狭屋
京都市中京区二条通小川角
電話 (075) 231-0616　FAX (075) 252-2020

松屋常磐
京都市中京区堺町通丸太町下ル
電話 (075) 231-2884

紫野源水
京都市北区北大路新町下ル西側
電話 (075) 451-8857

あとがき

私自身、よくもまあ、これだけ和菓子の写真を集めたものだとビックリしています。

本書は、私のブログ（京男雑記帳）がきっかけでできました。ブログを開始したのが2005年7月10日。ブログ、イコール日記と考え、日々感じたことを書き始めましたが、日記を毎日書くという作業が、ものぐさ者の私に続けられる訳がないと内心思っていました。

ブログのテーマは「京都人から見た京都」です。観光で来られる方には、わからない京都というものがたくさんあるので、それをテーマにしたら面白いのではと、その頃、仕事で京都中の「おはぎ」を食べ比べるという企画をしていました。いろいろ調べている時に京都の和菓子というものに行き当たったわけです。

小さい時から見慣れていた和菓子屋さんが突然気になり始めました。そのなかでもまず上生菓子といわれるジャンルのお菓子の造形的な美に驚かされました。以前フランス料理の学校に通って料理の研究や技術の習得をしたのですが、フランス料理にはない、さらには京都で生まれ育っていたにもかかわらず、和菓子のすばらしさに気付いていませんでした。

元禄時代に出版された『女重宝記』『男重宝記』という本があります。元禄時代の若者に紳士淑女のための教養を示した本です。『男重宝記』の巻之四にお菓子の項目があり、お菓子に関する蘊蓄が書いてあります。

紳士は、お菓子の名前・作り方・いただき方などを理解していないといけないとあります。上生菓子を「見て、名前を聞いて、味わい」そして「景色を感じる」という、それがブログのテーマになってきました。京都の社寺や行事が私にとって色鮮やかに見えてきました。

上生菓子との出会いは、いつも緊張します。きっとお作りになる菓子職人さんの気迫なんだと思います。私がたくさんある上生菓子から特定のものを選ぶのは、そういう気迫を感じているからかもしれません。

そういう上生菓子と出会うのは、縁なんだと思います。なぜなら上生菓子はいつ行っても同じものは置いていない。早いものは、次の日にはお作りになっていない場合だってある。出会いなんです。そして手に入れたら急いで撮影できる場所に帰らないといけない。その時、箱を揺らしてはいけない。漆の丸盛皿にのせる時も緊張します。

最初竹の菓子用のお箸がなかった時は、香道の時の銀の火筋(こじ)を使っていました。それを見てあんまりだということで、菓子職人さんが自分でつくる竹箸をいただきました。
また、素人の悲しさで上生菓子の中側がどうなっているのか知りたくなり、シタビラメを５枚におろす時の薄くて弾力のあるナイフで切りました。これも気合いが入ります。躊躇うと失敗します。きっと見る人が知りたいだろうな……と。これに関しては菓子職人さんに失礼なことをしているのかもしれません。この場を借りて謝ります。でも素人は知りたいのです。とくに外国から来られた人に「クリームはどんなのが入っているのか？」と聞かれることが多かったのです。
羽二重や外郎、わらび餅系を二つに切るのは、難しかったな……いまでもそうです。

本書をご覧いただいた方にいくつかお願いがあります。
本書に登場するお店にはいつも同じものが置いてある訳ではありません。季節や行事をテーマに刻々と変化していきます。同じお店でも職人さんによっても違います。入手した後は時間との勝負です。乾燥すると色がドンドン褪せていきます。だから遠方から観光で京都に来られてもお土産には不向きです。手に入れたらホテルや旅館に帰り、美味しいお茶で楽しむのがいいでしょう。

上生菓子との出会いを大切にしてください。茶道でいう「一期一会(いちごいちえ)」に通じるものだと思います。

最後に、本書を作成するにあたり、上生菓子の写真をこういう形で使ってもいいと寛大にお許しいただいた各店の店主さんたちに厚く御礼申し上げます。
つたないブログに来ていただいたみなさま、あなた方がおられなかったらきっと続けていなかったでしょう。これからもよろしくお願い申し上げます。
また、本書を作成するきっかけをいただいた藍風館の大前正則さん、デザインをしてくださった中西睦未さん、その他の関係各位に深く感謝申し上げます。

最後に、母方の祖父である山川久三郎さん、遺伝子をいただいたことに感謝します。

2013 年 1 月吉日
中村　肇

著者プロフィール

中村 肇 Hajime Nakamura

1952年 京都西陣で生まれる
大阪芸術大学短期大学卒業後、大阪芸術大学に編入。在学中に宣伝会議主催コピーライター養成講座修了。
日本調理師学校フランス料理・洋菓子専攻修了。京都の広告代理店、デザイン事務所勤務を経て独立。
中村ハジメ事務所を主宰し現在にいたる。デザイナー、商品プロデュース、コンサルタント、企業理念構築など
幅広く活躍。ブログ「京男雑記帳」は日本全国から多くの読者に支持されている。

Born in Nishijin, Kyoto 1952
After graduating from Osaka University of Arts Junior College, incorporated into the Osaka University of Arts.
Training course completion advertising copywriter while attending the University. Department of French cooking school Japan
completion pastry-cooks. Through independent advertising agency in Kyoto, the design office work. To today presided over
the office Hajime Nakamura. Designer, goods producer, consultants, and building corporate philosophy worked extensively.
"Kyo-otoko Zakki-cho" blog is supported by many readers from all over Japan.

装　丁: 釜内由紀江、石川幸彦 (GRiD)
本文レイアウト: 中西 睦未
翻　訳: ルーシー・マクレリー (ザ・ワード・ワークス)
企画・編集: 大前 正則 (藍風館)

和菓子

2013年1月30日　初版発行
2018年1月20日　新装版初版印刷
2018年1月30日　新装版初版発行

著　者　中村 肇

発行者　小野寺優
発行所　株式会社河出書房新社
〒151-0051 東京都渋谷区千駄ヶ谷2-32-2
電話 03-3404-1201(営業)
03-3404-8611(編集)
http://www.kawade.co.jp/

印刷・製本　大日本印刷株式会社

Printed in Japan
ISBN978-4-309-27914-5